Marcio Luis Fernandes

GUARATIBA AND ITS GEOGRAPHICAL AREAS

Marcio Luis Fernandes

GUARATIBA AND ITS GEOGRAPHICAL AREAS

Guaratiba places and domains in the Rio context

ScienciaScripts

Imprint

Any brand names and product names mentioned in this book are subject to trademark, brand or patent protection and are trademarks or registered trademarks of their respective holders. The use of brand names, product names, common names, trade names, product descriptions etc. even without a particular marking in this work is in no way to be construed to mean that such names may be regarded as unrestricted in respect of trademark and brand protection legislation and could thus be used by anyone.

Cover image: www.ingimage.com

This book is a translation from the original published under ISBN 978-620-6-76082-5.

Publisher:
Sciencia Scripts
is a trademark of
Dodo Books Indian Ocean Ltd. and OmniScriptum S.R.L publishing group

120 High Road, East Finchley, London, N2 9ED, United Kingdom
Str. Armeneasca 28/1, office 1, Chisinau MD-2012, Republic of Moldova, Europe
Printed at: see last page
ISBN: 978-620-7-76620-8

Copyright © Marcio Luis Fernandes
Copyright © 2024 Dodo Books Indian Ocean Ltd. and OmniScriptum S.R.L publishing group

*"Domains, limits, frontiers and obstacles are issues frequently
focused on in geographical studies" (MELLO, 2001, p. 89).*

JOÃO BAPTISTA FERREIRA DE MELLO: A STAR THAT NOW SHINES IN OTHER FIELDS

I met João Baptista Ferreira de Mello in person in 2001 during my degree in geography. On that occasion, he was coordinating a walking tour of Rio de Janeiro's city center as part of the Urban Geography course taught by Professor Alvaro Ferreira.

Twenty-three years ago, however, I was already familiar with the writings of this first-rate geographer, as well as his reputation as a star at UERJ. JBFM, as he called himself, is considered one of the main exponents of humanistic geography in Brazil, a current with which I identified due to the influence of his "carnivalesque" texts.

For this reason, as soon as I finished my undergraduate degree, I went to UERJ to be supervised by him in my specialization in territorial policies in the state of Rio de Janeiro, a course I completed in 2006. In 2008, I entered the master's program and once again chose João as my advisor. In 2012, the same happened during my doctorate. When he found out he was ill, João was very sorry that he couldn't advise me on my post-doctorate. Even so, he didn't give up on any of his students. He used to say: I'm not dead yet. When I die, you can find another supervisor.

João was called up (as he put it) on July 10, 2021. Better than having been his student and mentor for more than 15 years, it was enjoying his company for all that time. I will never forget his lectures, his good humor and his hilarious jokes. More than a great work, Professor João Baptista Ferreira de Mello leaves behind a generation of geographers of the first magnitude committed to a much more enlightened world.

JBFM has also left an immense void called saudade in the hearts of all of us. The feeling he aroused in his students and advisees is much more than admiration. His name will forever be engraved in our hearts and minds...

Marcio Luis Fernandes

Guaratiba, May 07, 2024.

DEDICATORY

I consecrate this book to my Good God, the source of all my inspiration. He is the Lord of everything and everyone, the owner of all things. He sent his only son, Jesus Christ, Lord and Finisher of my faith. He has been with me at all times, not allowing this work to become a solitary endeavor.

I also offer this book to my dear mother - Vandi Bastos de Moraes - who passed away during the course of this research in December 2023. Without the support of this incredible woman, I certainly wouldn't have made it this far. She has always been my greatest supporter and for this reason she deserves my affection, admiration and love.

I dedicate this work to the greatest gifts I have ever received from heaven. I'm talking about my daughter Nicole Bazzani Fernandes and my son Luigi Bazzani Fernandes, who were brought into existence during my career as a researcher. My life gained depth and a new meaning after the arrival of my little ones.

I couldn't conclude this dedication without mentioning my dear Guaratiba neighbors and friends. In fact, this investigative work was written by a resident of Guaratiba, considering his trajectory in the midst of this carioca domain.

ACKNOWLEDGMENTS

Before our long teacher-student relationship, Professor Miguel Ângelo Campos Ribeiro's fame had already spread throughout the geography department at the Moacyr Sreder Bastos University Center, the institution in Rio's West Zone where I began my geography studies. The presence of one of the greatest names in Brazilian geography at our college was always a source of honor, and the auditorium was always packed with undergraduates willing to listen to his creative approaches to diverse topics such as the interaction between tourism and geography, the territories of prostitution, reflections related to the space of Rio de Janeiro, among other topics linked to human geography.

Since my arrival at UERJ in 2005, Miguel's participation in my university career has been emblematic. I was always impressed by his teaching and his charisma in the classroom.

In fact, Miguel was present at all my exams at UERJ (specialization monograph, master's degree qualification, dissertation defense, doctoral qualification and the defense of my thesis), always contributing to my improvement. To my dear Professor Miguel I would like to express my sincere thanks, respect, affection and admiration.

PREFACE

GUARATIBA, A PLACE OF EXPERIENCE AND MEMORY, BY MÁRCIO LUIS FERNANDES

It's a pleasure for me to preface the book that Marcio Luis Fernandes is giving us, the result of his post-doc, from 2023 to 2024. The author of this book is a teacher, geographer, researcher and writer. Having been born in Ilha de Guaratiba, a neighborhood in the far west of Rio de Janeiro where he lives to this day, this born Guaratiba native has always carried his lived universe with him. On March 5, 2024, Guaratiba turned 445 years old and, following its urbanization march, with a process of expansion of the city of Rio to the West Zone, it has not lost its beauty and charm.

In 2005-2006, Marcio Luis Fernandes took a specialization course in Territorial Policies in the state of Rio de Janeiro, when I met him and he was my student. I sat on his exam boards for his specialization in 2006, his master's degree in 2010 and his doctorate in 2015. At the end of 2023, he began his post-doctoral internship under my supervision. Marcio follows the Humanist current of Geography, based on the approaches of his advisor, Professor João Baptista Ferreira de Mello, supported by various authors, including Yi-Fu Tuan. Based on these great thinkers, Marcio points out that "we are always influenced by the experiences we share in our lived universe" (Fernandes, 2024, p.141).

As the author of this book sees it, Ilha de Guaratiba is one of those small communities, proclaimed by residents like himself, a lived neighborhood that pulsating daily life has transformed into a center full of values, an inseparable whole, made up of people, friends, acquaintances, relatives, a territorial base, evocations and other references that allow those who experience it the pleasant sensation of feeling at home or even immersed in the center - navel - heart of the world, as Tuan and Mello pointed out in their elocubrations (Fernandes, 2024, p.142).

In fact, there are several Guaratibas covered in this new book by Marcio Luis Fernandes. It is not a homogeneous unit. In fact, different "Guaratibas" coexist: Guaratiba, Barra de Guaratiba, Pedra de Guaratiba and Ilha de Guaratiba - in a boundary that occupies part of the West Zone of the municipality of Rio de Janeiro and makes up the Guaratiba Administrative Region - made up of three official neighborhoods: Guaratiba, Barra de Guaratiba and Pedra de Guaratiba. Within Guaratiba, the largest neighborhood in the municipality of Rio de Janeiro, there is another neighborhood whose boundaries are fluid and existentially demarcated - which the municipality has yet to include on the municipal map: Ilha de Guaratiba.

This book is structured in three parts: the first deals with the memory of places. In the second part, it looks at the trajectory of a Guaratibano, from the 1970s to the 2020s. In the third part, he discusses Guaratiba and its domains. At the end of the book, Marcio, on page 144, points out that "for me, Guaratiba is the 'place par excellence', the cradle of my history and geography, the place has always welcomed me in my years of existence"... in my opinion, there is no place like Guaratiba.

Therefore, the book presented and, as already pointed out, working with the categories of the Humanist current of Geography, focuses on Guaratiba, a place where the work takes in its political, natural, symbolic and affective domains.

City of Rio de Janeiro, Laranjeiras, fall 2024.

Miguel Angelo Ribeiro.

SUMMARY

FIRST WORDS

Guaratiba is a place that can be used as an example of our city's expansion process. The peripheral portion of Rio de Janeiro's west zone is currently undergoing a veritable urbanization march, as its former rural-agricultural configuration has been replaced by today's urban-residential trend (FERNANDES, 2003; 2006; 2010; 2012; 2014b; 2015c; 2020).

The growing real estate speculation in the area began in the 1970s and intensified in the last years of the 1990s. This process has metamorphosed Guaratiba into one of the veins of Rio's urban sprawl in recent years. The event in question has stimulated considerable mobility towards the place (FERNANDES, 2020; 2021; 2024).

The bucolic location, visited sporadically by owners of second homes - a traditional agricultural producer - is undergoing a constant process of land/real estate appreciation and a considerable increase in its resident population. The area in question has for years been presented as the most likely target for volatile speculative real estate capital. Many experts point out that the city of Rio de Janeiro will grow towards Guaratiba (LESSA, 2001; REDONDO, 2012; JANOT, 2013). In the second half of the 16th century and the following century, Guaratiba was a large rural estate where sugar mills, oxen and horse farms and vegetable plots dominated. At the time, its expansion was rapid, thanks to the hard work of Velloso's descendants. By the end of the 18th century, many sugar mills were multiplying, including Engenho Novo, Engenho de Guaratiba, Engenho da Ilha and

Engenho do Morgado, which once belonged to Father João Pereira de Cerqueira and is now the Mudinhos still (FERNANDES, 2022; 2022b).

In order to transport the production, the mangrove swamp had to be cleared in order to support this demand. Pinto (1986) reports that some boats sank in the river when the sugar production from the Morgado sugar mill was being transported. The sugarcane plantations and sugar mills were joined in the 19th century by coffee plantations, grown on the hills of Ilha de Guaratiba (FRIDMAN, 1999).

In the context of the outskirts of the city, until the first half of the last century, Guaratiba represented the so-called "carioca hinterland", and its ruralism and rusticity was the focus of an important study by researcher Magalhães Corrêa in 1936 (CORRÊA, 1936). Because they live in a peripheral location, far from the central area, Guaratiba residents are often the target of jokes and mockery by those who live in the city's better-known neighborhoods (FERNANDES, 2019; 2020; 2021).

Until the 1990s, Guaratiba, it is worth repeating, was characterized as one of Rio's last rural remnants - an area where the production of fruit, vegetables and legumes prevailed. From this period onwards, however, the bucolic location, frequented sporadically by owners of second homes, underwent a constant process of land/real estate appreciation and a considerable increase in its resident population.

Secondary residence, or second home, is a category of analysis widely used by geographers to explain the urbanization process in states such as Rio de Janeiro (RIBEIRO E COELHO, 2007) and localities such as Guaratiba (FERNANDES, 2003; 2006; 2010).

The term "domain", in turn, can be understood as an existentially demarcated area that refers to a notion of belonging to the ground experienced by women and men amidst the singularities of their respective lived universes. This is a category commonly used in geographical studies (MELLO, 2001) and which, in the specific case of this book, represents the references (symbolic, natural, political, affective...) that surround the spatiality focused on in this study, which is Guaratiba (FERNANDES, 2022b).

The decline of agricultural production turned Guaratiba into an area of notorious real estate greed, as evidenced by the residential condominiums that sprang up after 1990. It is worth noting that most of the new residents migrated from neighborhoods in the South Zone and other locations such as Barra da Tijuca, Recreio dos Bandeirantes, Tijuca and Jacarepaguá in search of closer contact with nature (FERNANDES, 2008; 2009; 2010).

In fact, since the first centuries of colonization, landfills have contributed to Rio de Janeiro's urban expansion. As the city grew, man overcame elevations by drilling tunnels, such as the Rua Alice/Barão de Petrópolis in 1887 and the so-called Túnel Velho in 1892, connecting Botafogo to Copacabana. More recently, Barra da

Tijuca and Recreio dos Bandeirantes have gained prominence amid the sprawl of Rio's urban fabric (CARVALHO, 2004).

In the current race to the west, the aforementioned urbanization march continues to incorporate new areas into its valued space. With a view to the 2016 Olympics, the Grota Funda Tunnel, inaugurated in 2012 as part of the Transoeste corridor, definitively integrates Guaratiba into Rio's urban fabric (FERNANDES, 2011; 2012; 2013).

This synthesis represents a significant part of the approaches that we will try to decode through this book, the aim of which is to address the similarities and differences between the different guaratibas in the midst of the changes and continuities in their domains (FERNANDES, 2014b; 2014c; 2015c).

The research is justified by the need to focus on a strategic portion of the city of Rio de Janeiro (LESSA, 2001), a picture that is being rapidly modified by the growth of the city of Rio de Janeiro towards the Guaratibano domains (FERNANDES, 2014; 2015; 2016).

The area in question has been widely studied by agronomists, biologists, environmentalists and archaeologists who seek to highlight its importance in relation to fauna, flora and nature in general. However, the area in question lacks research that focuses on the human element in relation to spatial and environmental changes, a field that is specific to the social sciences and geography in particular (MORAES, 2007).

The author of this research has lived in Guaratiba since birth, and is a pioneer and connoisseur of its areas, having a strong identification with the place focused on in this book. Of course, this fact alone is not enough to justify the academic work that culminated in the publication of this book. However, Guaratiba is a relevant place to be explored and, under these conditions, to be the focus of studies and research, which is why the investigation in question was developed with greater motivation. Furthermore, my academic career is intertwined with the domains of Guaratiba, a place that has always welcomed me in my years of history and geography (FERNANDES, 2016; 2018; 2024).

On this path, while still an undergraduate, I began to develop my first research on my lived universe: Ilha de Guaratiba (FERNANDES, 2003). As the place was undergoing a real metamorphosis in its geographical domains, I continued to develop approaches linked to it in my specialization (FERNANDES, 2006), master's (FERNANDES, 2010) and doctorate (FERNANDES, 2015c).

Now, in the post-doctoral internship of which this book is the result, I have sought to broaden the geographical limits of my research, pursuing the geographies of other Guaratiba domains: Guaratiba (neighborhood), Barra de Guaratiba and Pedra de Guaratiba (FERNANDES, 2019b; 2020; 2021; 2022b).

In this context, the research aims to present an approach based on a framework of experiences, symbolic and affective experiences, among others, based on the author's understanding of the transformations that the place in question has been undergoing, sharpening his study based on the trajectory of a Guaratibano who built his history and geography in the Rio de Janeiro area studied here (FERNANDES 2023b; 2024).

The choice of the line of research (humanism in geography) is justified by the characteristics of the study area, specificities related to its agricultural past and the natural attributes that have always distinguished this part of the city of Rio de Janeiro. Furthermore, the spatial transformations linked to the urbanization march in Guaratiba have also produced considerable changes, both in its natural environment and in the lives and relationships of its residents. With this in mind, I believe the research topic is pertinent to the line of inquiry, since the relationship between people, their place and nature is at the heart of the research we intend to develop (FERNANDES, 2014; 2014b).

The aim of this research is to uncover the similarities and differences between the different Guaratibas, in other words, the singularities that strengthen the geographical ties between the localities in question and the particularities that distinguish the different neighborhoods in the Guaratiba Administrative Region. It's worth repeating that this investigation is based on my particular geographical understanding as a Guaratiba native who has always lived, worked and carried out research in this area of Rio de Janeiro (FERNANDES, 2010; 2011; 2012; 2013).

From the process of renewing geography and the emergence of critical geography - understood as the currents that broke with the fossilized truths of the positivist and neo-positivist ideals - geographers of various methodological orientations (existentialists, Marxists, eclectics, etc.) began to take on a more human perspective, seeking, through the transformation of the social order, a more generous geography and a fairer space (place), which is organized in function of men (MORAES, 2007).

In this sense, we think that the best way to focus on the human being at the center of all things (TUAN, 1982; 1983; 1998; MELLO, 1991; 2000; 2001) would be to use an individual's intimate and particular collection in approaches to their lived space, where their approach would be the basis of scientific investigation. In this way, a more vibrant, generous, authentic and - why not say it - more human geography would emerge, since the geography produced by those most interested in their lived reality can also produce a place based on their real needs and/or desires. The research consists of applying the premise described here to a geographical context that is the object of our investigation: the Guaratiba Administrative Region (FERNANDES, 2024).

Following this perspective, we will use some of the principles defended by hermeneutics and phenomenology, philosophies of meaning that understand place to be an integral part of being, with each individual being an informal geographer capable of discussing the soul of places, since it is man who produces, learns, lives and transmits geography (SCHUTZ, 1979; BUTTIMER, 1982; LOWENTHAL, 1982; COSGROVE, 2004, MELLO, 2004; 2005; 2007).

Because it focuses on an individual in their lived place, our research - it bears repeating - is based on the phenomenological method, while also using another philosophical support belonging to the niche of philosophies of meaning, namely hermeneutics. The theoretical bases for this research were sought in authors who are adept at humanism in geography and who are based on subjectivities, symbolism and identities, therefore, in researchers who use the phenomenological method and hermeneutics, such as Relph (1976), Buttimer (1982), Tuan (1982; 1983; 1998), Holzer (2001; 2008) and Mello (1990; 1991; 1993; 1999; 2000).

We intend to maintain the methodological line based on the assumptions of phenomenology and hermeneutics, which seek to decode existential geographies through qualitative analysis where the individual is important and their lived world fundamental (ARANHA, 1996; ABBAGNANO, 2007) amid the indivisibility of subject/object. Therefore, the proposed methodology is based on intense bibliographical research - the result of an accumulation of readings since graduation - and the collection of philosophical material that has provided us with relevant theoretical and conceptual support. Under these conditions, I turned to the experiences I've had in the midst of my introjections into Guaratiba's scenarios. In other words, my memorable geographies (FERNANDES, 2019; 2021; 2023) were the empirical basis of this investigation in an attempt to decipher the geography(ies) of Guaratiba's domains based on and committed to this lived world. Before writing a little about the trajectory responsible for the geographical construction that guides this work, I think it is pertinent to bring up a brief approach to the relationship between place and memory.

1   ON THE MEMORIES OF PLACES

As we have argued in previous research (FERNANDES 2010; 2015c; 2020; 2023), places are attributed many dimensions of meaning (BUTTIMER, 2015). One of the ways pointed out by different authors in order to explore the dynamic and multifaceted lived universe would be through a glimpse of the intimate and private collections of those who experience it (FERNANDES, 2020; 2021).

However, since every lived experience inexorably goes back to the past, if we want to follow this path, we will need to rely on the individual and collective memories of individuals and social groups (LOWENTHAL, 1982; 1985; 1998; ABREU, 1998; MELLO, 2002; HALBWACHS, 2013; BUTTIMER, 2015).

No matter how much a locality is transformed, as a result of urbanization, for example, it will always retain a residual content of past temporalities. There are, however, a range of events that leave no physical traces. These are often recorded in the geographical memory of the individuals who follow these changes (FERNANDES, 2023).

What we think and feel about our geographies is obviously based on experiences from somewhere in the past. From this link, what happens on our experienced ground becomes part of our very existence (LOWENTHAL, 1998).

Everyday life is based on place, which is the result of all our moments, the sum of our memories and the product and combination of all our lived experiences (MENDILOW, 1960). Possessed by the past, Lowenthal admits to living in a tangle of eras. For him, the past resides in us through our memories and recollections of different moments (LOWENTHAL, 1985; 1998).

By populating the thoughts of human beings, the past is alive in our memory (HIGHET, 1949). In fact, the scenarios and experiences glimpsed and lived through become part of our memory (BUTTERFIELD, 1965).

Thus, "all awareness of the past is based on memory. Through memories we regain awareness of previous events, distinguish between yesterday and today, and confirm that we have lived through a past" (LOWENTHAL, 1998, p.75).

Memory is both personal and collective. The memories of individuals and social groups sustain their sense of identity in relation to their experienced ground. Our lives are impregnated by our memories. At every moment, we bring back some event from the past (LOWENTHAL, 1998).

Memories tend to accumulate with age. Our stock of memories increases as life goes on and experiences multiply. Although there are individual and collective memories, memory is always personal. Memories cease to be personal only when we decide to share them (LOWENTHAL, 1998).

The memory of different spaces and places is inscribed not only in some artifacts from the past that persist in their landscape as geographical testimonies, but - above all - in the memories of the experiences lived by men and women in their lived universe. In this context, place transcends materiality, although it is not dissociated from it, because spatial memories are eternalized in our geographical memories (MELLO, 2002; FERNANDES, 2023).

As a common trait, individuals are excellent connoisseurs of the dynamism of their lived universe (BUTTIMER, 1982; 2015), and are therefore able to reveal the nuances of their affective and frequency domains more aptly. In this context, they resort to their geographical memories, since their lived experiences undoubtedly refer to their past experiences. In order to share something related to their place or bring back experiences they once had there, women and men resort to their memories, thus bringing these memorable geographies into the present (FERNANDES, 2010; 2015c; 2023b).

With regard to the relationship between the geographical concept of place and the memories of its inhabitants, let's consider the opinion of geographer João Baptista Ferreira de Mello, according to whom

> the concept of place, traditionally explored by geographical knowledge, is intertwined with the trajectory of the humanistic approach, taking on, within this horizon, the meaning of home, as it is full of experiences and resourcefulness, at the same time, a respite of stability and well-being, as well as a welcoming home and an introjected field of movement and belonging to be defended. These principles were learned and adapted from phenomenological philosophy and from the dynamism of the lived world, an inseparable whole made up of people, a territorial base, belongings, events, acquaintances and all sorts of elements that allow the individual to feel at home (...). In this whirlwind of permanence, understanding, exchanges, actions, conflicts, dreams, delusions and affection, our geographies are perpetuated, because place transcends materiality, although it is

not dissociated from it, because spatial memories are eternalized in our memories (...). Human geography is concerned with the organization of space. Geographers from the humanistic wing do not deny this perspective, reworking the concept of place based on the lived experiences, feelings and understanding of individuals and social groups, pointing out its multidimensionality and the various ways of understanding it (MELLO, 2002, p. 63).

As Mello points out in the quote above, the phenomenological notion of place that underpins humanistic thinking considers the individual and their territorial base to be inseparable. Places, as we know, are the result of distinct temporalities and spatialities (FERNANDES, 2014; 2016).

In order to achieve a better understanding of the place or to restore places from the past, the geographer often resorts to memories of the place's past. These memories may be embedded in a monument or artifact of some kind. On the other hand, the past of places or the places of the past may still be present in the memories and recollections of individuals and social groups who have their places as a support for their lived experiences (FERNANDES, 2020; 2021).

In another line of geographical thought, Professor Maurício de Almeida Abreu (1998, p. 82), defines the term memory as the "capacity to store and preserve information", being "an essential element of a place's identity". By bringing together different individual memories, the memory of a place becomes a collective memory. In this sense, the subjectivity introjected into individual memory can converge into intersubjectivity, since different memories can converge to create a shared memory. We can see, then, the premise that the individual memory of the residents of a neighborhood, for example, can contribute to the recovery of their memorable geographies or their geographical memories (LOWENTHAL, 1982; 1998; ABREU, 1998).

Let's proceed by appreciating the assumption that geography is everywhere and that each individual can be considered an informal geographer, able to discern the multifaceted universe they experience (LOWENTHAL, 1982; COSGROVE, 2004). Based on this premise, through the records contained in this individual's geographical memory, it is possible to walk the paths of their memories and uncover geographies of different temporalities, as well as past spatialities and artifacts that have been lost in the vortex of time, spatially speaking, but remain alive and pulsating in the memory of the human beings who experienced those moments. The importance of this recovery for the identity of a place is unquestionable. In the midst of a country that tends to neglect its past, oral histories (and geographies), as well as the memories of old people (and adults and young people, too), are in the process of being disseminated in Brazil (ABREU, 1998; BOSI, 2003).

The experiences lived by individuals in a locality give rise to memories that, although very different from each other, have in common the adherence to the same place. Pursuing the necessary recovery of "time in place", or "the set of temporalities" (ABREU, 1998, p. 94) that makes it possible to unveil its past and present geographies (FERNANDES, 2010), we seek to elucidate the ideas related to the construction of a geography based on memory. After all, "it is impossible to deal with the empirical without coming to it with a previous theoretical background" (ABREU, 1998, p. 88).

From these elucidations to others, having traveled the path so far with the purpose of providing a philosophical basis for our approach, considering the experience and understanding of the author of this research who is also a resident of Guaratiba, we will now move on to the exercise concerning the fusion of stories x philosophy. In the light of humanistic thinking, which does not distinguish the individual from their place, we will then use the geographic memory of a Guaratiba resident to unveil the geographies of this Carioca domain.

2   THE TRAJECTORY OF A GUARATIBANO

My childhood was spent among guava trees, orange trees and banana trees, in a very Brazilian and Rio de Janeiro style, but with the distinctive touches of a peri-urban suburb. The games included flying kites, pique-flag races, pelada and marbles. As my father, Manoel Fernandes, was a gravedigger, I was obliged to take his lunch to the Ilha de Guaratiba Cemetery. That's why I had no feelings of rejection for the field of the dead. As a teenager, because I had few resources, I started working as a market vendor selling vegetables that I had grown myself. All this led me to an experience of strong attachment to my lived universe and its people, internalized within me.

On this journey, intertwined with my place, I use my lived experience and, together with the members of my community, I undertake this saga about our lived universe. This memorable place, inscribed in our hearts and minds, Guaratiba, has been metamorphosed by ambition and speculative perspectives on its domains. I was born here. I saw my neighborhood with its rural/agricultural physiognomy, in previous decades, now replaced more strongly by the residential urban with a view to becoming a spatial portion of enormous economic value of its soil, on behalf of real estate agents counting on the encouragement of the Public Power. All my life, I've lived in the same neighborhoods as the rest of the Guaratiba Administrative Region. Over the course of my five decades of lived experience, I have followed the changes and continuities that have marked the places I have lived, as well as the similarities and differences between the different Guaratibas. I would like to share these geographical memories, starting with my childhood and adolescence in the 1970s and 1980s.

2.1   **The 1970s and** 1980s

My childhood takes me back to a completely different place. Over the last five decades (since 1970), Guaratiba has undergone a real spatial metamorphosis. Its productive structure has changed, its landscape has changed, the way people deal with each other has changed - in short - a lot has changed. I remember, as if it were today, the green plantations that abounded in my street. The lettuce, cabbage, watercress, parsley, chives and chicory gardens - as well as the okra, chayote, pumpkin, jiló, sweet potato and cassava fields (IMAGE 1) - were many. Not to mention the mango, orange, papaya, passion fruit and tangerine orchards.

Like a considerable portion of Guaratibanos with a long tradition, I lived through a period when access to sizable plots of land was still relatively easy due to the human vacuum that the area represented at the time. In the meantime, citrus farming was practiced by some local farmers. One of Magalhães Corrêa's descriptions of the area in question is "the region of orange groves" (CORRÊA, 1936). This fact indicates that the production of oranges predates the beginning of my trajectory in my lived universe.

IMAGE 1 - THE FIELDS OF AGRICULTURE - A manioc plantation I photographed on Ilha de Guaratiba (FERNANDES, 2010).

With regard to the occurrence of orange groves in Rio de Janeiro, let's freely consider the elucidations of Abreu (2008). The researcher points out that citrus production acted as a major brake on the subdivision wave until the first half of the last century, preventing some municipalities in the Baixada Fluminense and neighborhoods in the traditional West Zone of Rio de Janeiro from being hit by the real estate fever of the time.

However, with the outbreak of the world conflict, exports collapsed, as all the oranges were exported on foreign refrigerated ships that no longer docked in Rio de Janeiro. In addition, the lack of refrigerated warehouses and poor road transportation from the farms to the railroad led to the fruit rotting on the trees, creating a citrus plague that would decimate a large part of the plantations.

At the end of the First World War, in 1945, with production no longer serving the domestic market, the export of oranges was banned, dealing the coup de grace to those who had managed to preserve their orange groves during the crisis. From then on, the orange groves were replaced by allotments in municipalities like Nova Iguaçú and in neighborhoods like Campo Grande - major orange producers in that period (FERNANDES, 2015c). This urbanization march, however, did not take place in Guaratiba, since many of its orange groves were kept to meet internal demand and the rest were replaced by a diversification of agricultural crops that came to characterize the place as a true green belt - a major producer of fruit and vegetables.

The rusticity and rurality of the "sertão carioca", with its inland aspect and its farmlands, already impressed Rio de Janeiro's urban "outsiders" in the 1930s (CORRÊA, 1936). Can we imagine this fascination 50 years later, when the city had already taken on the airs of a metropolis? It was surprising to most cariocas that, within their own city, in a metropolis like Rio de Janeiro, there was a place completely different from the world they lived in. A place characterized by farmland and a life devoted to nature and rustic countryside (IMAGE 2).

IMAGE 2 - THE DOMINIONS OF PECUARIA - The countryside that still predominates in Guaratiba (FERNANDES, 2003).

Agricultural activity in Guaratiba was closely related to the world of street markets (IMAGE 3). In fact, agriculture in the area was maintained for decades thanks to the street markets. Until the 1980s, most Guaratibanos had two jobs: farmers and market traders.

IMAGE 3 - THE DOMINIONS OF THE FREE MARKETS - One of Campo Grande's free markets on a Sunday morning. For decades, this type of popular market was the economic mainstay of rural producers in Guaratiba (FERNANDES, 2006).

When CEASA was built, it had a direct impact on local agricultural production. Many farmers who grew the produce they sold at the street markets started buying their produce at CEASA. Then the local sacolões sprang up, which directly affected the street markets. Most people stopped going to the markets when sacolões opened in their neighborhood. In addition, the prices at the sacolões were more attractive.

I remember well the big market traders and producers in the area: Mr. Nori, a big banana producer; Mr. Manoel Fonseca, who produced vegetables; Mr. Alfredo, a big producer of chayote and passion fruit. I myself was a market trader and farmer in my teens.

29

Another factor that contributed to the ruin of Guaratibano farmers was the competition from large producers in the Serrana region and the interior of São Paulo. Gradually, the local farmers realized that there was no way they could compete with the products coming from abroad. Many of them replaced their fields with the production of ornamental plants.

In addition, by becoming a supply center that receives products from various municipalities in the state, CEASA began to compete with rural producers in Guaratiba, since many market traders began to buy products in this market that, in the context of the city of Rio de Janeiro, were only found in the far west of Rio de Janeiro. By competing internally with food cultivation from 1990 onwards, the production of ornamental plants - because it represents a much more lucrative activity - also contributed to the decline of the traditional agricultural activity in Guaratiba.

According to Mascarenhas (1991, p. 1), "the street market consists of a periodic group of retailers" who display their merchandise in versatile structures, using the public highway for this purpose, and therefore depending on a concession from the municipality to temporarily take over the street (p. 12). Therefore,

> the market vendor does not own the space he uses, unlike traditional commerce. They only acquire, on a temporary basis, the right to use that space on days of the week and at times pre-established by the government, to display certain products (MACARENHAS, 1991, p. 13).

Widely dispersed throughout the city of Rio de Janeiro, the street market has played an important role in urban supply over time, especially in the food sector (vegetables, fruit and fish). From the elite neighborhoods of the South Zone to the suburbs of the West Zone, this periodic gathering is part of everyday social life in Rio.

From the 1970s onwards, however, a new type of retailing came onto the scene: supermarkets. Their rapid expansion in the city ushered in a period of strong competition with the traditional periodical markets, compromising their performance and radically changing their spatial distribution (MASCARENHAS, 1991).

With regard to Guaratiba's past spatiality, we can see that the aura of the place favored agriculture and that most of the arable land was used by its owners - notably small and medium-sized farmers - to produce fruit, vegetables and legumes. These products were largely sold at street markets located in different neighborhoods. Local agriculture was geared towards this retail trade, providing local producers with a market for their produce and supporting this activity in Guaratiba for many decades.

In fact, the breakdown of the agricultural system that had always characterized Guaratiba was due to a series of events, including the gradual decline of the street markets. The genesis and subsequent proliferation of the sacolões throughout the city's neighborhoods contributed to this, which dealt a severe blow to the street markets and, by extension, to the local farmers who depended on the dynamism of this retail market.

The opening of the fruit and vegetable supply center (CEASA) in the Irajá district also had a negative impact on local agriculture. Many market traders opted to buy directly from CEASA and abandoned their fields. After its inauguration (in 1974), CEASA became a direct competitor of Guaratiba's rural producers.

According to Mascarenhas (1991), the decline of free markets began with the advent of supermarkets in the 1970s. This decline was aggravated by the proliferation of popular greengrocers (sacolões) throughout the city from 1990 onwards (FERNANDES, 2006), leaving the majority of agricultural producers in Guaratiba without a consumer market for their products. The loss of dynamism of the open-air markets meant that traditional agricultural activity collapsed dramatically in the area and forced it to be restructured.

Despite the transformations brought about by the decline of its traditional agricultural activity, some traces of other times still remain in the area. An example of this are the farmers who are still trying to hold on today, persisting with their traditional vegetable plantations. Other producers, however, have started to grow ornamental plants on the land previously used for agriculture.

In fact, the production of ornamental plants competed internally with agriculture, since it became a much more lucrative activity than food production from 1990 onwards. This allows us to uncover part of the past geographical context of the place, related to agriculture and its relationship with street markets.

This Guaratibano who is writing to you would also like to point out the dawn of new activities linked to the production of ornamental plants, gardening and landscaping, branches that have replaced traditional agricultural production since the 1990s, as well as the appearance of the first residential condominiums in the area.

2.2 **The 1990s and 2000s**

The influence of the prestigious landscaper Roberto Burle Marx on the process that led Guaratiba to migrate from traditional agriculture to ornamental production is indisputable. From the 1990s onwards, Burle Marx's many works began to demand a large quantity of ornamentals, which his production site could no longer handle. As a result, his employees began to support this production by growing plants on their properties, which had previously been used for agriculture. From then on, these new flower growers went on to become owners of their own businesses and influenced many to change their industry and their lives. But it all started with the encouragement of Roberto Burle Marx.

Roberto Burle Marx was struck down by abdominal cancer in June 1994, at the age of 84. When he died, he left behind 2,000 designed gardens. Born in São Paulo and moved with his family to Rio in 1913, he grew up in an environment surrounded by greenery in Leme, where from an early age he helped his mother cultivate species in the family garden.

A student of architecture and painting, the future landscaper worked as a visual artist until his neighbor Lúcio Costa, who admired the plants he cultivated, invited him to do the landscaping for one of his projects. This is how, in 1932, Roberto Burle Marx signed his first garden - for the Schwartz family home - in Copacabana, making the 23-year-old famous.

Burle Marx's last work was a project for Kuala Lumpur, Malaysia, which was on his drawing board when he died and was completed by his office team (CALS, 1995; SÁ, 2008).

In order to cope with the large number of plants that his many gardens required, in 1949 the landscaper acquired the old Santo Antônio da Bica site - now the Roberto Burle Marx site - with more than 35,000 square meters, located at the foot of the Serra Geral de Guaratiba, where he began to produce, decorate and collect hundreds of ornamental species (IMAGE 4).

Due to the many friendships he made and the great affinity he had with the place, as well as the need to be closer to his work, Burle Marx moved permanently to Guaratiba in 1973, thus intensifying his influence on the place (CALS, 1995; SÁ, 2008).

In the same way that fruit and vegetable production replaced citrus farming after its crisis in the 1940s and 1950s, guaranteeing the agricultural aptitude of Guaratibanos, floriculture - especially after 1990 - has been replacing the traditional gardens of Guaratiba. This phenomenon, however, would not have been so successful were it not for the prestige of Roberto Burle Marx, who represented a milestone, a watershed between one economic segmentation (agriculture) and another (landscaping).

IMAGE 4 - THE ROBERTO BURLE MARX SITE - Partial view of the Roberto Burle Marx Site in Guaratiba - in the far west of Rio de Janeiro (FERNANDES, 2008).

In this respect, we can say that a new geography emerged in Guaratiba as a result of the unfolding of Roberto Burle Marx's life and work. The oldest gardens in the area began to gain visibility after the eminent landscaper began to spread Guaratiba throughout the city. Today, the Guaratiba areas are a major producer of ornamental plants, known even outside the state of Rio de Janeiro, thanks to the prestige of this artist.

As well as providing the local community with a viable alternative to work, Burle Marx is one of the main references in the area and is constantly remembered. An example of this is the adoption of his name by the largest school in the area and the replacement of the name of the old Barra de Guaratiba road, which was renamed Roberto Burle Marx Road.

Being passionate about nature and the bucolic nature of its landscapes, Burle Marx - who was also a painter - chose Guaratiba to live, work and produce his canvases, using its beautiful scenery as a backdrop and inspiration. This is how this eclectic artist produced hundreds of paintings, magnetized by the scenic beauty of the place he chose as an anchor and also to live his experiences (TUAN, 1980; 1983; SÁ, 2008).

Despite maintaining the land's primary function, the production of ornamental plants is committed to a new structure that is emerging in the area, related to the real estate market. Currently, its main function is to supply the growing demand from condominiums and construction companies, whose "fetish landscaping" is one of their strategies for transforming land into a real estate commodity.
In Guaratiba, however, this speculative process began in the 1970s with the acquisition of farms and plots of land that gave rise to the first second homes in the area, making Guaratiba a popular place for those wishing to escape the hustle and bustle of the metropolitan area for a while.

Among the reasons that have led many families to buy a second home in Guaratiba are its natural attributes, its bucolic atmosphere and its tranquillity - as opposed to the hustle and bustle of the metropolitan area.

If we relate the phenomenon of the acquisition of second homes to the analysis of Rio de Janeiro's urbanization process, we cannot ignore the role of speculative real estate capital in the housing production process, which is one of the main factors responsible for the sprawl of Rio's urban fabric (ABREU, 2008; RIBEIRO, 1997).

In this context, Ribeiro and Coelho (2007) point to the emergence of new ways of living aimed by real estate capital at the most affluent classes, which have contributed to the spatial restructuring and expansion of metropolises. Residential condominiums are one example of this.

However, this logic related to the production of urban space often begins with the process of acquisition of secondary residences by the economically privileged classes, as Assis (2003) points out for a general context and Fernandes (2003; 2006; 2009) for a specific context, based on the process of urbanization in Guaratiba.

In his research on the spatial expression of the phenomenon of second homes, geographer Lenilton Francisco de Assis is concerned with trying to elucidate the main causes of the spatial event in question.

In his analysis of the spatial repercussions of the phenomenon of second homes, Assis (2003) points out that, as a result of the metropolization of certain cities, it became increasingly necessary for urban people to leave the overcrowded central areas for the metropolitan peripheries in search of a reunion with nature.

This was a way of relieving their daily stresses and renewing their energy. The urban space, which was once the center of attraction for housing and rural people in search of work, now, despite concentrating various functions, leads its residents to seek out new areas that offer them the necessary conditions for using their free time in contact with nature.

Therefore, due to their proximity to the central areas, the metropolitan peripheries became the main targets of real estate speculators who sought to valorize the natural and cultural attributes of these spaces, offering them to specific social segments that had surplus income to purchase a secondary residence.

Assis (2003) also proposes that the phenomenon of second homes is one of the factors responsible for the process of urbanization of the periphery, since certain capital migrates to peripheral areas, materializing through properties that also come to represent a reserve of real estate value.

By assuming the availability of a surplus income, the second home is no longer just a leisure alternative, but also an investment option. As soon as the second home also has an exchange value, the person most responsible for transforming the peri-urban space comes into play: the real estate speculator. The latter, through insidious advertising, aims to transform the natural attributes and amenities of the metropolitan periphery into real residential attractions, as we can see below in relation to real estate speculation in Guaratiba:

> Finally, it's your turn to live in paradise! Right here on Earth. Skunks and prey in droves cross the asphalt suspiciously. Oxen walk leisurely in single file towards the pastures. Marmosets in droves monkey around in the branches of the trees. Herons and wild ducks, in their lightness, develop scenic choreographies over lakes and streams in spectacular flights. That's life around here. Time seems to have stood still. It's definitely the new Recreio! The difference is that here you can still negotiate with decent caipiras at low prices. Mini sites for you to live near the good and the best, "far and near" expensive and saturated areas (FERNANDES, 2006, p.42).

In the quote above, taken from an advertising folder (J. Brandão Negócios Imobiliários), what strikes us most is the emphasis placed on the natural attributes of the place, a fact that contributes to its real estate plots being offered as part of a true paradise lost in the middle of the metropolis of Rio de Janeiro. This real estate advertisement from the early 1980s reveals the genesis of a practice that is very common today, but with a very significant difference: the romanticism that the advertisement is trying to convey is justified by the situation of the place at the time.

At that time, Guaratiba was nothing more than a town characterized by rural activities - mainly based on agriculture and the many farms visited by their owners during their leisure time. It was the typical weekend in the countryside. As such, the target audience for this type of advertisement was people interested in acquiring a relatively large area where they could build a farm which, in most cases, also housed a secondary dwelling. However, from the early 1990s onwards, new real estate agencies began to offer the economically privileged a product that differed from that offered by its precursors. Thus, secondary housing gradually began to be replaced by horizontal condominiums where housing became permanent.

Returning to Assis' (2003) approach when he looks at the process of transforming second homes into permanent residences, the geographer explains that in the last phase of this process, the original perimeter is absorbed by metropolitan expansion, now forming part of the city itself, while the previous second homes are metamorphosed into permanent residences.

Today, more and more people in Guaratiba are choosing to live in their old secondary homes. In addition, families coming mainly from the Recreio-Barra-Zona Sul axis, have found a satisfactory quality of life in the area, considering its locational advantages, such as relatively low prices, availability of land, landscape attributes, low crime rates, among others. The latter, however, represent a new type of resident: those who opt for horizontal condominiums.

Despite symbolizing a new spatial dynamic in Guaratiba, the phenomenon of housing production - on a broader scale of analysis - reveals, as Ribeiro (1997, p.199) points out, the continuity of the process of urban expansion in the city of Rio de Janeiro:

> From the second half of the 19th century, especially after 1870, the city of Rio de Janeiro underwent important urban transformations generated by the action of a group of capitals that began to invest in urban space. Among them, what we might call real estate capital, applied to the production of rental housing and the purchase, subdivision and sale of plots of land previously used for agricultural purposes...

In the specific case of Guaratiba, the dismantling of the rural-agricultural structure, prior to the process of consolidating the place as a reserve of real estate value, triggered a process of spatial transformation that has been changing its spatial appearance.

Considering that each location combines variables from different times (SANTOS, 1997), in the place we studied, we noticed the existence of elements representative of phases prior to the process of change in question that still resist amid the innovations. As examples, we can mention some remnants of rural activities from the past and the persistence of many sites that are still used as a leisure option during vacations and weekends (second homes).

Since process is the continuous action that implies time, continuity and change (SANTOS, 1992), we understand that the phenomenon of second homes represented the genesis of the metamorphosis that Guaratiba is undergoing. Thus, in the midst of the second homes, in the early 1990s, the first residential condominiums began to appear, as a result of the growing real estate development in the area.

In addition to the natural attributes and the country atmosphere, other advantages cannot be ruled out if we want to understand the urbanization march that is taking place in Guaratibanos. Associated with these factors is the lower tax burden (IPTU) compared to Recreio dos Bandeirantes, Barra da Tijuca and neighborhoods in the South Zone. In addition, the proximity factor can also be included in this package, since Guaratiba is relatively close to the aforementioned locations.

Although it has a country atmosphere and a rustic ambience, Guaratiba represents a Carioca domain. This allows new residents to be in a place that meets their needs, paying a much lower price compared to Barra and Recreio, neighboring districts.

As a resident who knows my locality very well, I would like to summarize here the characteristics of the area, which - in my opinion - are preponderant for the real estate development that culminated in the construction of the current residential condominiums. In the meantime, in addition to the factors already mentioned, I would like to highlight the availability of land, as well as the relative proximity of the central area and neighborhoods with commercial and service sub-centers, as the main factors in real estate development that promoted both the aforementioned residential mobility and the consequent genesis of the condominiums that were built to cater for this new and growing public.

Created mainly in new areas of the metropolis with natural amenities, the exclusive horizontal condominium is the result of a process of effective land appreciation and real estate development. Becoming the "eldorado" for an upper middle class that comes, in part, from the old noble areas of the city, condominiums are characterized by the self-segregation of social groups who, having an income, can live wherever they want. And the choice of new residence is influenced by the massive advertising surrounding the amenities and the new lifestyle (CORRÊA, 1992).

In the 1990s, the first residential condominiums sprang up in Guaratiba as a result of the growing value of real estate. The area, which for centuries had been little explored and which locals said had stood still in relation to other places in the municipality of Rio de Janeiro, then began to attract the attention of many people in the midst of countless metropolitan problems.

Its domains began to be an object of desire when it came to living in a place full of amenities, offering those wishing to escape the metropolis a quality of life compatible with their desires. In this sense, it is worth repeating that the attributes of the place, such as mountains, greenery, tranquillity and safety, among others, are then used as attractions for those who wish to be, paradoxically, close to and far from urban life.

As explained above, the genesis of the race towards the so-called "lost paradise" in the interior of the metropolis occurred when several second-home owners began to live there permanently. This phenomenon caught the attention of some real estate speculators, who began to exploit the charms of the location in their advertising.

This type of speculative activism persisted throughout the first half of the 1990s, thus increasing the demand for real estate on a scale where the old properties, such as farms and mini farms, could no longer cope. The first horizontal residential condominiums then began to appear, built on the flat areas of the site, especially on the tidal plain, through intense landfilling of mangroves and alluvial channels.

Ever since the first condominiums were built in the Guaratiba lowlands, real estate has been increasing in value. With each passing day, living in places like Ilha de Guaratiba becomes more expensive and difficult, especially in the condominiums which, with very few exceptions, represent an option only for people from upscale areas like Recreio dos Bandeirantes, Barra da Tijuca and neighborhoods in the South Zone.

This fact is justified by the high value placed on land and/or real estate, which is only accessible to the economically privileged classes who, for the most part, live in highly valued areas of the city.

As it is a type of development that requires a lot of land, and is built on areas of environmental preservation, horizontal residential condominiums, as well as representing the pillar of a new urban-residential trend, are the most responsible for the reproduction of countless environmental impacts. Since the infrastructure that could better support the urbanization process in the area studied is non-existent, environmental degradation becomes a constant concern for its residents.

2.3 **The 2010s and** 2020s

The current urban trend in Guaratiba has not only metamorphosed the place, but has also caused a series of environmental impacts in its geographical areas. Thus, by achieving their goal of bringing many families from other parts of the city to the area, real estate developers have also fostered a real spatial metamorphosis, since this mobility presupposes structural changes to accommodate it. In this context, nature gradually gave way to real estate developments that promoted a series of environmental impacts, a problem rejected by Guaratibanos who, like me, are not willing to give up the physical attributes that characterize our lived universe.

With the march of urbanization that characterizes Guaratiba's domains, the increase in the number of residents has also led to an increase in the amount of garbage and contamination of the rivers, which used to be populated by lobsters, crabs and fish. In the forest we also found sloths, anteaters, capybaras and other animals. Today, however, much of this wealth no longer exists, and there is still the risk of losing what is left of our exuberant nature.

The downside of this whole process of real estate development and urbanization is the lack of basic infrastructure. Without the necessary urban planning, this growth can pose a huge risk to the best thing about the place: its nature. The lack of sanitary sewage has contaminated rivers and groundwater. Deforestation on the hillsides and the landfilling of mangroves have been constant, further contributing to environmental degradation.

With regard to the feelings and sensations of individuals and social groups in relation to an attack on their lived world, we can deduce, based on Tuan (1980), that environmental problems are fundamentally human problems. This thinker is certainly based on a totalizing vision, also shared by Corrêa (1992), which includes the human element in its conceptualization of the environment. However, both in the midst of introjections and in relation to the approach of human geography that privileges human beings, including them in its attacks, it is also undeniable that every human action on nature (re)produces environmental impacts (DREW, 2002).

Expanding the scale of analysis around the relationship between urban growth and the impacts caused to nature, it is interesting to refer to the emblematic process of consolidation of the metropolis of Rio de Janeiro which, compressed between the sea and the mountains, flanked by beaches, sandbanks, marshy lowlands and forests, had its growth forged in the struggle for space and in overcoming the distances generated by that same growth, in the face of the special conditions of its physical environment (GALVÃO, 1992).

> Squeezed between the mountains and the sea, Rio de Janeiro had in these natural elements the great beacons of its expansion (...).
> The development of technology allowed these obstacles to be gradually overcome, enabling the city to incorporate into its built fabric spaces that were once considered unsuitable or unlikely for urban occupation (ABREU, 1992, p.54).

Among the spaces considered unsuitable or unlikely for the sprawl of Rio's urban network were, in addition to the massifs and mountains mentioned by Abreu, also alluvial channels, sandbank areas, as well as mangroves and forests. All these old limiting factors, however, were no obstacle to metropolitan growth. Thus, from the first centuries of colonization, landfills contributed to the organization of Rio's urban space. As the city grew, the elevations had to be broken up by drilling tunnels since 1887 (CARVALHO, 2002). More recently, Barra da

Tijuca and Recreio dos Bandeirantes have gained prominence amid the sprawl of Rio's urban fabric. In recent decades, Rio has grown towards the Guaratiba plain (FERNANDES, 2015b), amid fears among its residents about the consequent deterioration of the natural biosystems that characterize the area.

Concerns about Guaratiba's natural domains are justified because, as a lowland area with ample and exuberant greenery, largely surrounded by the Pedra Branca Massif (Serra Geral de Guaratiba), Guaratiba's most important feature is its beautiful natural landscape, which is the backdrop used by real estate agents to attract followers of a lifestyle based on closer contact with nature. The area is basically made up of two conservation units, namely the Guaratiba Biological and Archaeological Reserve, located on the tidal plain, and the Pedra Branca State Park, of which the Serra Geral de Guaratiba hills are also part (ATLAS DAS UNIDADES DE CONSERVAÇÃO DA NATUREZA DO ESTADO DO RIO DE JANEIRO, 1990). In addition, according to the Rural Union and Rio de Janeiro City Hall, Guaratiba represents an important area of rural and environmental preservation. This information, as mentioned, is ratified by the city council itself, in a municipality considered to be urban.

Geographically speaking, every human activity advocates environmental transformations, since the environment - or space - presupposes, in addition to natural and built material forms, also the human being who gives it dynamism (SANTOS, 2002). Therefore, the more unprepared an area is in terms of basic infrastructure to support the desired changes, the greater the environmental impacts will be.

This is the case in Guaratiba. As if the changes represented by the influx of new residents and the construction of housing were not enough, the lack of infrastructure in the area further increases the damage to nature.

Considering only the actions promoted by the phenomenon described by Abreu (1992; 2008) as "real estate fever", in recent decades the place has undergone the first external interventions through the installation of secondary residences. In the meantime, deforestation has increased on the slopes of the Serra Geral de Guaratiba and also in some isolated areas of the lowlands in order to make way for the sites that housed the houses used during vacations and weekends. Many of the streams that flowed down the massif, bringing crystal-clear water to the flat areas, were dammed upstream to form lakes and natural pools on the new properties. These rivers, which until then had been populated by fish - such as traíras, caraúnas, piabas etc. - and crustaceans such as lobsters and pitus (large freshwater shrimp) are gradually losing their fishiness.

Subsequently, the increase in the inflow led to an even greater impact, since many of the old streams began to receive untreated sewage, turning them into veritable black ditches. As this process unfolded, many hills were demolished, contributing to land and real estate speculation, giving way to new homes and serving as landfill for new subdivisions in an area that until then had not suffered a major impact: the tidal plain.

It was from the 1990s onwards that environmental degradation took off. The constant landfilling of mangrove areas and alluvial channels, with the aim of building condominiums, caused an unprecedented impact on the mangrove biosystem, which resulted in a sharp decline in the population of crabs and guaiamus, which, as well as being considered one of the symbols of the place, also represent a source of income for many families of pickers.

Due to the lack of planning and structural basis, the spatial changes discussed here do not represent a de facto new urban structure, but rather a radical change which, in my opinion, is one of the main factors responsible for the degradation of the natural setting of the place. This phenomenon, however, by changing the spatiality in question, can also change people's lives, since the influences between people and their places are reciprocal.

Extending this process of spatial transformation even further, this time to a global analysis, the growth of cities and the urbanization of the world is, according to Wirth (1976), one of the most remarkable facts of modern times. The social scientist recalls that the change from a rural society to a predominantly urban one took place in the space of a single generation in some central countries. The author, however, is also concerned with the difference between urbanism and urbanization, where urbanization refers to the urban form, represented by buildings, and urbanism, to the peculiar lifestyle of cities. Many other theorists have also addressed the issue of the distinction between the city and the urban, since common sense usually refers to these terms as synonyms.

Because it represents a hyper-complex reality and has several dimensions, the urban environment is represented in different ways. In this sense, Yázigi (2003), in his discourse on the various values surrounding the urban environment (historical value, social value, economic value, affective value, etc.), stresses the importance of a sense of belonging in building a permanent urban heritage. According to Yázigi, without a sense of belonging you can't expect much from an urban agglomeration, which is increasingly becoming a modern form of camping. Affective value, therefore, is an indispensable condition for building an urban space that is understood as heritage and as a topophilic environment (TUAN, 1980).

For Corrêa (2000), the need for greater consumption of space as a result of the increase in the value of real estate in the central areas, at the same time points to much more advantageous conditions in the distant peripheries, which are endowed with amenities. Under these circumstances, the process of urbanization of a peripheral part of the city of Rio de Janeiro like Ilha de Guaratiba, much more than a mere spatial change, represents subjective transformations in which its residents come to experience a different lifestyle, based on new values. In this sense:

> Urbanization no longer denotes merely the process by which people are drawn to a place called a city and incorporated into its system of life. It also refers to that cumulative accentuation of the characteristics that distinguish the way of life associated with the growth of cities and, finally, with the changes in meaning of the ways of life recognized as urban (WIRTH, 1976, p. 93).

For Tuan (1980, p. 260) "the suburb is an ideal, because it suggests a perfect lifestyle, in which the best of rural and urban life are combined without their defects". In pursuit of this alternative lifestyle, middle-class people have been moving to Guaratiba. Because of the violence and crime, families with favorable financial conditions tend to look for quieter places to live. Although it's no longer the same as it was a few decades ago, the place still has that playful rural feel and, for now, it continues to breathe that atmosphere, but I don't know for how long. That's what makes Guaratiba different in 2023.

(Sub)urban areas like Guaratiba represent an ideal for certain individuals and social groups. Among the reasons that led many families to move from traditional neighborhoods in Rio de Janeiro to the peripheral area in question was the search for a quieter place to live. A place where a new lifestyle in closer contact with nature and the countryside could flourish without having to completely leave the city behind.

Although he comes close to phenomenology and hermeneutics in some of his research and conferences, geographer João Rua does not share humanism in geography. However, in order to elucidate the issue related to the multiple interactions between rural and urban areas, we believe his assertion is pertinent. In his research, the aforementioned thinker (RUA, 2002) emphasizes both the issue of the manifestations of the urban in the rural, a phenomenon he calls urbanities, and the strength of the rural in the face of the urban (ruralities). For Rua, the new ruralities command the current process of spatial restructuring, since the rural, when incorporated into the general process of urbanization, is integrated into the urban while retaining some specificities.

In his theoretical considerations about urbanities and new ruralities, Rua moves towards approaches that refer to the hybridity of these new spatial realities (SANTOS, 2002), since the two realities (con)merge. Thus, the "old" or "internal", represented by Guaratiba's past rural structure, when associated with the "new" or "external", portrayed by its current urbanization process (SANTOS, 1997), give the place a range of new characteristics, specificities, particularities, singularities, as well as a broad niche of recent and past symbols that present themselves to our interpretation.

3   GUARATIBA AND ITS DOMAINS

For many outsiders who hear about Guaratiba, it represents nothing more than a peripheral area far from everything and everyone. However, because the passion they experience doesn't share their dulled thinking, their ethnocentric attitude towards geography, Guaratibanos like me place their place at the center of their world.

Representing a common human tendency, ethnocentrism leads people to overvalue their lived universe, elevating it to the status of the best place in the world. This is the positive side of ethnocentrism. However, this attitude also has negative aspects. In this sense, people who live in the city's upscale neighborhoods tend to discriminate against residents of areas they consider less important (TUAN, 1986; 2013; MELLO, 1991).

But what is Guaratiba? How big is it in Rio's territory? What are the neighborhoods in this region of Rio de Janeiro? What unites the different neighborhoods of Guaratiba? What differentiates them? What are the characteristics that distinguish the different localities in Guaratiba? What are the symbols and/or symbolic geographies of these spaces and/or places in the understanding of their residents, or even of a resident who - phenomenologically speaking - connects with their geographical domains? These are the questions to be answered in the light of my geographical understanding, which is based on a deep relationship with the domains of this area in the far west of our city (FERNANDES, 2021).

3.1 **Guaratiba before Guaratiba**

The geographical construction of Guaratiba has always been linked to the arrival of the Portuguese in their dominions from 1579. This information, however, is easily disputed given the existence of sambaquis with the discovery of human bones dating back approximately 2000 years. Sambaqui is a word of indigenous origin, whose etymology goes back to two elements of the Tupi language: "tamba", meaning molluscs, and "ki", meaning heap or deposit (ARAÚJO, 1987; MENEZES, 2005; FERNANDES, 2010).

Thus, sambaquis are piles of organic materials made up basically of mollusc shells and crustacean carapaces, bones and wooden artifacts and were formed by people who mainly inhabited the Atlantic coast between five and seven thousand years ago (ARAÚJO, 1987; MENEZES, 2005; FERNANDES, 2015c).

These records were made by archaeologists who have been researching the sambaquis in the area where the Army Technology Center is located since the 1970s. The artifacts discovered and the results of the research carried out are in the collection of the National Museum and the UFRJ library, respectively. It was in one of these spaces, the Zé Espinho Sambaqui, that the oldest inhabitant of Guaratiba was found, who lived around 2000 years ago. He was 38 years old and between 1.40m and 1.50m tall. It fed on fish and crustaceans and had strong teeth, as can be seen from the absence of cavities even at that time. It used canoes to get around the region's many lagoons and mangrove swamps (IMAGE 5). After reconstruction work in 2018, we were able to see what the face of this individual would have looked like, which was named "Ernesto" after one of the team's researchers (ARAÚJO, 1987; MENEZES, 2005; FERNANDES, 2017).

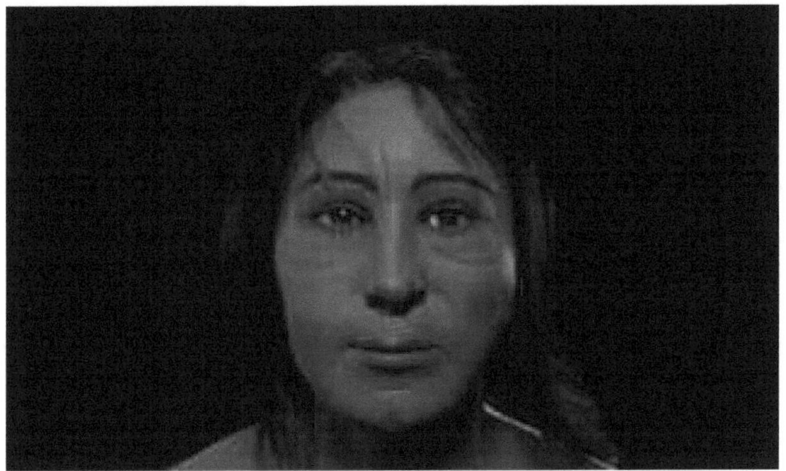

IMAGE 5 - THE PRE-HISTORIC MAN OF GUARATIBA - the first carioca in history and geography was from Guaratiba, having lived in its domains over 2000 years ago (FERNANDES 2015b).

Before the establishment of the Guaratiba sesmaria, the aforementioned area - as was the case with practically the entire Brazilian coastline at the time - was inhabited by the Tupis-Guaranis Indians, probable descendants of the "prehistoric man of Guaratiba". The indigenous people who lived in Guaratiba before colonization were the Tupinambás (IMAGE 6), who had inhabited the area for approximately 2000 years before the Portuguese arrived (ARAÚJO, 1987; MENEZES, 2005; FERNANDES, 2018).

IMAGE 6 - Illustration depicting the Tupinambá presence in Rio de Janeiro when the Portuguese arrived there in the 16th century. In: https://freewalkertours.com/pt-br/historia-rio-de-janeiro/. Accessed on 04/01/2024.

The place name Guaratiba is of Tupinambá origin: guará is a Tupi word meaning red and tiba, abundance. Etymologically, therefore, Guaratiba means abundance of guarás - red birds that populated the local mangroves in Tupinambá times (IMAGE 7) and which disappeared from the area in the 1980s (FERNANDES, 2003; 2006; 2010; 2015c; 2017; 2018; 2019; 2021).

IMAGE 7 - Guará: Guaratiba's symbol bird, responsible for the place's toponym (abundance of guarás). In: https://oglobo.globo.com/rio/lenda-em-guaratiba-guara-pode-voltar-para-casa-23275075 . Accessed on 04/01/2024.

The place name "Guaratiba" was therefore derived from the large number of wading birds that populated the area - the guarás. As the word "tiba", in Tupi-Guarani, means abundance, Guaratiba, etymologically, means "abundance of guarás" (PINTO, 1986). In these terms, the place name Guaratiba comes from indigenous words.

3.2 The policy areas

The long geographical construction of Guaratiba dates back to March 5, 1579, when Manuel Veloso received from the Portuguese Crown a plot of land measuring approximately 52 km² on which to settle (IMAGE 8). We are talking about the old sesmaria of Guaratiba, created 14 years after the foundation of the city of São Sebastião do Rio de Janeiro (FRIDMAN, 1999).

IMAGE 8 - THE DOMINIONS OF THE GUARATIBA SESMARY - The image above focuses on the Guaratiba Sesmaria. Founded on March 5, 1579, the boundaries of the area on canvas coincide with the domains of today's R. A. de Guaratiba. A. de Guaratiba. It is therefore the "original Guaratiba" (FERNANDES, 2020).

The arrival of Manoel Velloso, who came to live in the newly constituted sesmaria of Guaratiba with the other members of his family, was the political milestone of the Guaratiba domains. From then on, the area began to be exploited economically through the

58

construction and management of sugar and brandy mills for export (PINTO, 1986; FERNANDES, 2023).

But what were the sesmarias? When were these political units instituted? What was their purpose? Sesmarias were plots of land belonging to Portugal and handed over for occupation, first in Portuguese territory and later in its American colony (Brazil) where they were in force from 1530 to 1822. The word sesmaria derives from sesmar, to divide (PINTO, 1986; FERNANDES, 2022).

In Brazil, the sesmarias system was applied as a way of guaranteeing possession of the territory, which was already divided into Hereditary Captaincies. The captaincies guaranteed ownership and did not represent any expense for the Crown, but the territories suffered from invasions. The first distributions of sesmarias were promoted by Martim Afonso de Souza and consisted of the subdivision of the captaincies. The system guaranteed the colonization support needed by the Crown (PINTO, 1986; FERNANDES, 2021).

The distribution of land was aimed at attracting Christian settlers, who were guaranteed the right to use it through letters of donation. Those who received the sesmaria would not, however, have full administrative control and remained subject to the Crown. Among the main problems faced by the Crown in regulating the sesmarias was the obligation to cultivate and the establishment of territorial limits, which were often disobeyed by the squatters (PINTO, 1986; FERNANDES, 2020).

The squatters, to whom the sesmeiros leased the land, began to cultivate it and demand recognition of their rights over the territories. The Crown made numerous attempts to regulate the problem and it was only in 1822 that the sesmarias system was abolished, benefiting the squatters (PINTO, 1986; FERNANDES, 2019).

Over the centuries, sugarcane and coffee plantations, as well as other agricultural activities, left their mark on Guaratiba until the second half of the 20th century. The great agricultural production of yesteryear made Guaratiba one of the richest and most prosperous parishes in old Rio in the 18th and 19th centuries, until a great drought in 1888 consumed its plantations (SANTOS, 1965; FRIDMAN, 1999).

Image 9 focuses on the map of the administrative division of Rio de Janeiro in 1900, which privileged the political organization of Rio's territory into parishes. But what were the freguesias, the political units that existed in Rio in the past? Freguesia is the name given to the area of influence of a parish and, by extension, to the group of parishioners. With the close link between political power and that of the Catholic Church in Brazil before the Republic, the city of Rio de Janeiro was administered by parishes (SANTOS, 1965; FRIDMAN, 1999).

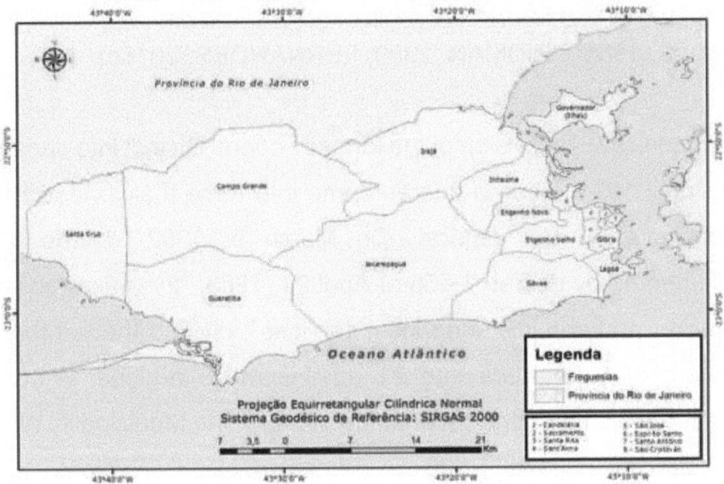

IMAGE 9 - THE DOMINIONS OF THE FREGUESY OF GUARATIBA - The image above focuses on the parishes of the city of Rio de Janeiro in 1900. Guaratiba was one of the largest and most important parishes in old Rio, especially between the 18th and 19th centuries. In: https://www.researchgate.net/figure/Mapa-da-divisao-administrativas-de-freguesias-do-Rio-de-Janeiro-em-1900-32-Evolucaofig 3295861217. Accessed on 10/01/2024.

In old Rio, the suburban parishes were Nossa Senhora de Apresentação de Irajá (1647), São Tiago de Inhaúma (1743) and Engenho Novo (1873); and those in the hinterland were Nossa Senhora do Loreto and Santo Antônio (Jacarepaguá, 1661); São Salvador do Mundo de Guaratiba (1755), Nossa Senhora do Desterro de Campo Grande (1673) and Santa Cruz - 1833 (SANTOS, 1965; FRIDMAN, 1999).

The localities that retain the name "Freguesia", such as those in Jacarepaguá and Ilha do Governador, have this name because they are the site of the parish church. In Irajá, the area around the Church of Nossa Senhora da Apresentação was also known as "Freguesia" (SANTOS, 1965; FRIDMAN, 1999; FERNANDES, 2015c).

The administrative division of the former Federal District into parishes lasted until 1934, when a decree came into force that divided Rio's territory into 36 tax districts. On March 9, 1962, decree 898, supplemented by decree 1.656 of April 24, 1963, "for the purpose of organizing and administering local services", divided the territory of the State of Guanabara into 21 administrative regions. With the subsequent dismemberment of some of them, the Municipality of Rio de Janeiro had 33 Administrative Regions by 2011 (FERNANDES, 2015c; 2024).

As we've seen, the old parish of Guaratiba was called "Freguesia São Salvador do Mundo de Guaratiba" (Parish of Saint Savior of the World of Guaratiba) - based in the current Parish of Saint Savior of the World. In other words, the aforementioned church was the parish church of Guaratiba. For this reason, the emblematic building is still known today as the Igreja Matriz de Guaratiba (IMAGE 10).

IMAGE 10 - TEMPLE OF THE GUARATIBA CHURCH - The Parish of São Salvador do Mundo is still known today as the Guaratiba Church. This is because this emblematic building represented the seat of the parish of Guaratiba (FERNANDES, 2022).

Currently, the political-administrative organization of the municipality of Rio de Janeiro favours the subdivision of Rio's territory into neighbourhoods. In fact, the city is made up of nine sub-prefectures that are subdivided into 33 administrative regions (which are equivalent to districts in most Brazilian municipalities). The Administrative Regions (IMAGE 11), in turn, are made up of one or more neighborhoods (FERNANDES, 2015c). The official division of the city of Rio de Janeiro into neighborhoods only took place in 1981, through Decree No. 3158 of July 23, 1981, based on the conceptual basis of the geographical notion of neighborhood defended by a renowned researcher: Professor Maria Terezinha Segadas Soares of UFRJ (SOARES, 1990).

IMAGE 11 - THE DOMAINS OF THE GUARATIBA ADMINISTRATIVE REGION - The XXVI R.A. represents Guaratiba, and is made up of three official neighborhoods (Guaratiba, Pedra de Guaratiba and Barra de Guaratiba) and a centenary neighborhood - so elected by its residents - whose city hall does not recognize as such: Ilha de Guaratiba (FERNANDES, 2020).

This study was carried out by Soares in 1962, within the framework of classical geography (SOARES, 1962), based on the approach of French geographer Pierre Monbeig (1957), according to whom a city is a collection of neighborhoods, each of which has its own physiognomy, resulting from the function of its inhabitants and their age. All these neighborhoods, more or less integrated with each other, make up the city. For Monbeig, an urban neighborhood has a character that belongs to it alone, a particular life, a soul. Influenced by this notion, Soares (1990, p. 105) lists:

> the notion of neighborhood is of popular origin, taken from everyday language. For the inhabitant of a city, the neighborhood constitutes, within the city, a whole that has

In zoning the city into neighborhoods, Soares (1962; 1990) considers the role of elevations to be important, since they can separate parts of the city, isolating them from the others. For the researcher, this physical characteristic can contribute to a place acquiring the individuality of a neighborhood. In this approach, the elements that could contribute to the individualization of a neighbourhood - in addition to the relief - would be the social content, the urban landscape and the function. The forces of the past and the factors of the present also contribute to the constitution of a neighborhood. Often, however, the municipal administration does not use this criterion, disregarding neighborhoods authentically elected by their residents, as is the case with Ilha de Guaratiba (IMAGES 12, 13 AND 14).

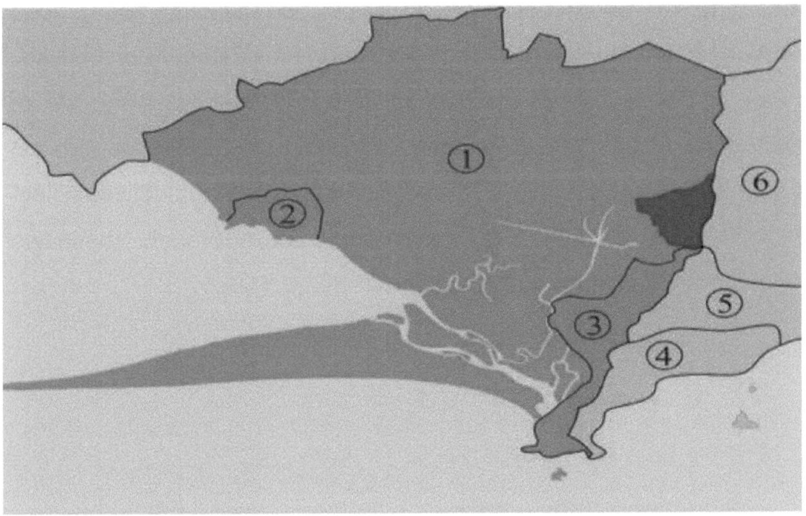

Caption:

→ Guaratiba Island (its domains until the 1990s)

→ (1) Guaratiba

→ (2) Pedra de Guaratiba

→ (3) Barra de Guaratiba

→ (4) Grumari

→ (5) Recreio dos Bandeirantes

→ (6) Vargem Grande

IMAGE 12 - THE LIMITS OF ILHA DE GUARATIBA UNTIL THE 1990S - The domains of Ilha de Guaratiba until the 1990s: location and surrounding neighborhoods (FERNANDES, 2003).

According to researcher Marcelo Lopes de Souza (1989), who focuses on the concept of the neighborhood as a category of analysis, it cannot be defined by neglecting its individuality and the life of relationships that make it dynamic. In this sense, "any neighborhood is simultaneously an objective and subjective/intersubjective reality" (SOUZA, 1989, p. 148). In its phenomenological approach, "these two dimensions interpenetrate" (SOUZA, 1989, p. 148).

IMAGE 13 - THE CURRENT LIMITS OF ILHA DE GUARATIBA - The limits of Ilha de Guaratiba (in red) today, after the real estate development process that is taking place in its domains (FERNANDES, 2023).

Like any place, the neighborhood is lived and experienced by its residents, making them (residents and neighborhood) an inseparable entity (SCHUTZ, 1979; TUAN, 1980; 1983; 2012; 2013; MELLO, 1991; 2000; FERNANDES, 2010; 2014; HALLEY, 2014). In his proposal for a holistic vision, dialoguing with humanism in geography, Souza (1989, p. 149-150) points out:

> The neighborhood has an identity that is intersubjectively accepted by its residents (...). A phenomenological look at the constitution of neighborhoods shows that the neighborhood corresponds to a certain part of the city which, by virtue of social relations, constitutes a lived and

67

felt space for the individual. The recognition and feeling towards the neighborhood comes from the fact that it is the place where we were born, where our house is, our friends' houses, the square we frequent (...). This range of feelings is essential to a neighborhood's identity.

IMAGE 14 - GUARATIBA ISLAND IN THE CARIOCA CONTEXT - The domains of Guaratiba Island (in green) in the Rio context (FERNANDES, 2020).

Although he points to intersubjectivity as an essential element in characterizing a neighbourhood, Souza (1989) does not ignore its objective dimension. For him, it is imperative that the two dimensions that make up the neighborhood interact in order to awaken in the residents a feeling of belonging to their neighborhood that the geographer, based on Tuan (1980), calls neighborhoodphilia (SOUZA, 1989, p. 151).

The current Guaratiba domains coincide with the boundaries established by decree 3158 of 1881 - the document that established the R. A. de Guaratiba - which is made up of three official neighborhoods (Guaratiba, Barra de Guaratiba and Pedra de Guaratiba). In addition to the three neighbourhoods mentioned above, within Guaratiba (the city's largest neighbourhood), there is another neighbourhood elected by its residents, which the city council does not recognize as such, whose boundaries are fluid and existentially demarcated: Ilha de Guaratiba. The "Guaratibas", therefore, are represented by a broad region within the Rio territory (FERNANDES 2003; 2006; 2009; 2010; 2015c; 2017; 2018; 2019; 2020; 2021; 2022; 2023).

3.3 Natural areas

The particularities of a given area are expressed, among other things, by the diversity of its natural elements. These physical attributes help us to reveal the geography and history of a place, since they represent elements on which its geographical construction takes place (FERNANDES, 2006, 2010; 2015c).

Like any other area, Guaratiba has specific physical characteristics that distinguish it from other areas in Rio (IMAGE 15). In this context, its natural attributes are clear examples of its particularity. Being a lowland area with ample greenery

IMAGE 15 - THE DOMINIONS OF SERRA GERAL DE GUARATIBA - The Serra Geral de Guaratiba is an elevation belonging to the "Pedra Branca Massif" that rises on Barra de Guaratiba beach, representing a natural boundary between the Guaratiba domains - to the west - and neighborhoods such as Recreio dos Bandeirantes (Photo by REDONDO, 2012).

and exuberant, surrounded in large part by the Pedra Branca massif (Guaratiba general mountain range), Guaratiba's most important features are its beautiful natural landscape, also made up of natural areas such as the Restinga da Marambaia, Guaratiba/Sepetiba Bay, mangroves, wild beaches and others (IMAGE 16).

IMAGE 16 - THE DOMINIONS OF THE ATLANTIC FOREST - On the slope of the Serra Geral de Guaratiba facing the sea - in Barra de Guaratiba - the green of the Atlantic Forest blends with the elevations of the Pedra Branca Massif to make up a memorable landscape (FERNANDES, 2005).

Relief is one of the fundamental elements of the physical area and its shapes and layers are studied by geographers. Because it has mountains and plains in its domain, Guaratiba has been the focus of research by these specialists. Over time, the local morphology has had a great influence on the construction of its history and geography. Its topography is basically made up of plain areas - where some "barriers" appear in the form of isolated rounded reliefs - and the elevations belonging to the Pedra Branca Massif. In Guaratibana's

71

morphology, the fusion between the vegetation of the lowlands, slopes and massifs provides a graceful view, both from the top of the massifs, from where we can glimpse a rich panorama, and from the plains themselves, where when we look at the elevations, we are faced with the picture presented by the hills that make its landscape even more breathtaking.

The Pedra Branca massif - spilling out over the sea - followed by escarpments, mountains and forested grottos - surrounds a considerable part of the Guaratiba lowlands, making it a kind of "island", surrounded by greenery and hills on all sides. One of the toponymic versions of one of Guaratiba's domains (Ilha de Guaratiba) even refers to the fact that the place is surrounded by the Serra Geral de Guaratiba like an island, limited by the elevations belonging to the Pedra Branca Massif.

The natural isolation caused by the Serra Geral de Guaratiba over the centuries has made access to this Carioca domain difficult. Like a natural obstacle, these elevations have partially isolated Guaratiba from other parts of the city, preserving its main natural features. When it comes to green areas, Guaratiba can be considered privileged (IMAGE 17). There are still important areas of environmental preservation and significant stretches of Atlantic forest (Fernandes, 2003).

IMAGE 17 - THE DOMINIONS OF THE WILD BEACHES - The only wild beaches in the city of Rio de Janeiro are located on the slope of the Serra Geral de Guaratiba facing the sea - in Barra de Guaratiba (FERNANDES, 2008).

In Barra de Guaratiba, at the foot of the Pedra Branca massif on its slope facing the Atlantic Ocean, there are five wild beaches: Praia dos Búzios, Praia do Perigoso, Praia do Meio, Praia do Canto and Praia do Inferno. These are known as wild beaches because of their difficult access and also because of the absence of buildings. In fact, this natural domain of Guaratiba can only be accessed by sea (by boat or speedboat), by air (by helicopter) or on foot. In this case, it is necessary to cross the entire Serra Geral de Guaratiba on a long walk through its steep elevations (IMAGE 18).

IMAGE 18 - THE TELÉGRAFO STONE - The Pedra Branca State Park - in Barra de Guaratiba - is home to the Pedra do Telégrafo, one of the most visited natural viewpoints in Rio de Janeiro (FERNANDES, 2022c).

In a recent survey, published by Veja Rio magazine on July 7, 2022, Pedra do Telégrafo was ranked among the 25 best nature walks in the world by an important tourism platform (https://vejario .abril.com.br/cidade/pedra-do-telegrafo-melhores-passeios-mundo/?Fb_clid=IwAR0IfzxLhHEUIenkAHPnFfkITxFK5dzg7TXAH5a EEYAfw1xE_Q3z-XNdX5CQ).

Located in the Pedra Branca State Park, more specifically in Barra de Guaratiba, the 354-meter-high peak stands out for the photos taken "on the edge of the cliff". The view is breathtaking and the images at Pedra do Telégrafo give the visitor the feeling of being on the edge of an abyss - reasons that, in themselves, justify the success of its trail. The 3.5-kilometer route, of light to moderate difficulty, makes up for the effort with the beauty of the surrounding fauna and flora and the privileged view of the beaches of Meio, Funda, Grumari and Restinga da Marambaia (FERNANDES, 2021).

The Restinga da Marambaia is a narrow layer of sand more than 40 kilometers long that separates the open sea from Sepetiba Bay. It stretches from Barra de Guaratiba to the municipality of Mangaratiba, where it forms Marambaia Island. For centuries, Marambaia served as a base for the landing of slaves who were forcibly brought from Africa to be traded. To this day, some descendants of these slaves live on Marambaia Island in quilombola communities (IMAGES 19 AND 20).

Located in the Pedra Branca State Park, more specifically in Barra de Guaratiba, the 354-meter-high peak stands out for the photos taken "on the edge of the cliff". The view is breathtaking and the images at Pedra do Telégrafo give the visitor the feeling of being on the edge of an abyss - reasons that, in themselves, justify the success of its trail. The 3.5-kilometer route, of light to moderate difficulty, makes up for the effort with the beauty of the surrounding fauna and flora and the privileged view of the beaches of Meio, Funda, Grumari and Restinga da Marambaia (FERNANDES, 2021).

The Restinga da Marambaia is a narrow layer of sand more than 40 kilometers long that separates the open sea from Sepetiba Bay. It stretches from Barra de Guaratiba to the municipality of Mangaratiba, where it forms Marambaia Island. For centuries, Marambaia served as a base for the landing of slaves who were forcibly brought from Africa to be traded. To this day, some descendants of these slaves live on Marambaia Island in quilombola communities (FERNANDES, 2023).

Although there are earlier reports, the first official document of ownership of Marambaia was registered in 1856 in the name of Commander Breves. In 1908, Marambaia officially became part of the Brazilian Navy. Today, the sandbank is a proving ground for the armed forces that administer this paradise. Dominated by extreme natural beauty, Restinga da Marambaia enchants those who have the opportunity to visit it (FERNANDES, 2021).

IMAGE 20 - THE LIMITS OF THE MARAMBAIA RESTINGA - The contours of the Marambaia Restinga (FERNANDES, 2021)

With an area of approximately 305 square kilometers, Sepetiba Bay is separated from the Atlantic Ocean by the Restinga da Marambaia (PICTURES 21 AND 22). In fact, because this body of saline and brackish waters is formed in and extends over the Guaratiba domains, it should be called Guaratiba Bay. Communicating with the Atlantic Ocean via two passes, its perimeter is approximately 130 kilometers.

IMAGE 21 - THE GUARATIBAN SHORE - Dusk in Pedra de Guaratiba on the shores of Sepetiba Bay (FERNANDES, 2021).

Covering approximately 3,360 hectares, the Guaratiba State Biological Reserve (RBG) protects an important mangrove remnant in the Metropolitan Region of Rio de Janeiro, associated with Sepetiba Bay. It is an ecosystem of great environmental, economic and social value, offering numerous environmental services, including

IMAGE 22 - SEPETIBA BAY - Covering some 440 square kilometers, Sepetiba Bay is bordered by the districts of Guaratiba and Restinga da Marambaia. In: https://www.facebook.com/Geografias Guaratibanas. Accessed on 01/24/2024.

These include maintaining biological diversity; providing resting and feeding points for various species of migratory birds; preventing flooding; as well as serving as a source of organic matter for adjacent waters, forming the basis of the trophic (food) chain for species of economic and ecological importance. Mangrove forests account for 1,601.34 hectares of the Unit's area; hypersaline plains or apicuns cover around 704.10 hectares, as well as wetlands and altered areas at different stages of regeneration (PICTURE 23).

IMAGE 23 - THE DOMINIONS OF THE MANGUEZAL - Located in the domains of the Guaratiba Biological Reserve, the Guaratiba mangrove is a true oasis in Rio. In: https://bafafa.com.br/mais-coisas/sustentabilidade/reserva-biologica-de-guaratiba-ecossistema-intocado-e-protegido-pela-unesco . Accessed on 01/24/2024.

The vegetation, made up of different species of mangrove, has a high degree of structural heterogeneity, with a great wealth of plant physiognomies. The Reserve was created by State Decree No. 7,549 of November 20, 1974, with the aim of preserving mangroves and archaeological sites, and was initially called the Guaratiba Biological and Archaeological Reserve. The RBG is also an integral part of the Atlantic Forest Biosphere Reserve declared by Unesco in 1992. It is also part of the Serra do Mar Biodiversity Corridor and the Carioca Mosaic, the latter created by Ministerial Order 245 of July 11, 2011 (FERNANDES, 2003; 2005; 2006).

Until the 1980s, Barra de Guaratiba beach was little visited compared to today. In that scenario, the Guaratiba beachfront was only frequented by bathers from the surrounding areas (Barra de

Guaratiba and Ilha de Guaratiba) and, as everyone knew each other, the coastline in question was affectionately nicknamed "hi beach" (IMAGES 24, 25 AND 26).

IMAGE 24 - THE BEACHES - Barra de Guaratiba beach on a winter's day. In: https://www.facebook.com/GeografiasGuaratibanas. Accessed on 01/24/2024.

However, from the 1990s onwards - with the environmental degradation of the beaches of Sepetiba Bay (Pedra de Guaratiba, Brisa, Sepetiba...), the cariocas of the West Zone began to have fewer leisure options on holidays and weekends, choosing Barra de Guaratiba beach as their point of interest (FERNANDES, 2021).

IMAGE 25 - THE DOMAINS OF BARRA DE GUARATIBA BEACH ON A TYPICAL SUMMER'S DAY. In: https://www.facebook.com/GeografiasGuaratibanas. Accessed on 01/24/2024.

Thus, the beach, which used to be frequented only by Guaratibanos from the island and Barra and was always empty, started to be "invaded" by thousands of bathers from the different neighborhoods of Rio's West Zone. It's worth noting that most of the bathers mentioned in this text use public transportation (bus 867 from the Campo Grande bus station). There is, however, another considerable movement towards the many restaurants in the area, this time by private car.

IMAGE 26 - THE DOMAINS OF BARRA DE GUARATIBA IN THE 1960S. The image above focuses on Barra de Guaratiba beach in 1962. In addition to the beautiful landscape of the beach in the 1960s, the stunning DKV Vemag that stands out in the colorized photograph on the GuarAntiga page stands out. In: https://www.facebook.com/GeografiasGuaratibanas. Accessed on 01/24/2024.

Barra de Guaratiba (IMAGE 27) is practically impassable on holidays and weekends - especially in summer - and its beaches are always crowded with bathers in the hottest months of the year.

IMAGE 27 - THE DOMAINS OF A GUARATIBA NEIGHBORHOOD. The image above partially focuses on Barra de Guaratiba, a neighborhood forged on the rocks of the Serra Geral de Guaratiba, an elevation belonging to the Pedra Branca Massif. In: https://www.facebook.com/GeografiasGuaratibana. Accessed on 01/24/2024.

Image 28 reveals part of the landscape of one of Rio de Janeiro's most beautiful neighborhoods. The photo of the residential part of Restinga da Marambaia shows Barra de Guaratiba in the background - a neighborhood that was literally forged between the sea and the mountains. In fact, Barra de Guaratiba has no flat land. As such, the houses in this residential and tourist neighborhood were built at the foot of the Serra Geral de Guaratiba, an elevation belonging to the Pedra Branca Massif that rises in this neighborhood with its disconcerting landscapes.

IMAGE 28 - BARRA DE GUARATIBA AND ITS NATURAL DOMAINS. In addition to the human constructions that merge with the elevations in the background, the image also shows the channel that connects Sepetiba Bay to the sea and the beginning of Restinga da Marambaia. In: https://www.facebook.com/Geografias Guaratibanas. Accessed on 01/24/2024.

The first human evocations in relation to nature remind us of fear and aversion to a hostile environment, where anthropic vulnerability was evident in the face of a wild habitat in which man was notoriously unfit to live (PARK, 1976; TUAN, 2005). The first cities and/or human settlements such as Jericho (in Palestine) and Ur in ancient Mesopotamia (now Iraq) arose millennia before Christ and were protected from enemy armies and the many dangers of nature by great walls (SOUZA, 2005).

It took the emergence of the great cities at the time of Alexander the Great for there to be a strong reaction in favor of the rusticity of natural environments. In these steps, when a society reaches a certain level of development and complexity, people begin to observe and appreciate the relative simplicity of nature. This kind of feeling only emerged after the construction of cities, when the pressures of urban life made rural peace and a romantic appreciation of nature attractive (TUAN, 2012). Nature, once a symbol of a topophobic environment (TUAN, 2005), nowadays represents an indispensable attribute for valuing areas in the intra-urban sphere and an essential merit for the pleasantness of places.

In this respect, the tripod of amenities, or just one of its items - sea-green-mountain - nowadays constitutes an element of valorization in large cities (CORRÊA, 2000; ASSIS, 2003; MELLO, 2007; ABREU, 2008). The scenery of ennobled areas, boasted by the wealth of green sea-mountain amenities, today represents a symbol of living well and of status in general. In Guaratiba, endowed with the aforementioned natural amenities, these predicates are part of the memory of those who are rooted in the place and, more recently, they attract and adjust to the expectations of the new residents as symbols of this carioca domain.

3.4 **The symbolic domains**

A symbol is a part that has the power to suggest a whole, transcends its condition as such and is confused with the place in which it is found (MELLO, 2000). Symbolism, understood as the expression or interpretation of the meaning of a given symbolic element (symbol), has emerged in recent decades as an extremely important concept for humanistic and cultural studies, studies related to understanding the subjective dimension of what is experienced (COSGROVE, 2004). Place - because it has a symbolic identity, charge, character and fervor - is full of symbolism, established by elements (material or immaterial) that evoke countless meanings to the territorial base experienced (MELLO, 2003). The place itself is a symbol of affection, well-wishing, satisfaction, happiness and togetherness (TUAN, 2012). The symbolic character of places establishes connections by decoding and translating their past and connecting it to the present. Considering the premise that places and symbols acquire deep meaning through the emotional bonds woven over time, reconciling, understanding and decoding the symbolic content of different places are tasks to be undertaken by geographers who are interested in the perspectives of the relationship between individuals and their lived universe (CORRÊA, 2000; FERNANDES, 2010; 2015c).

Places are full of symbols, and this precept is defended by geographers of the humanistic horizon, according to whom places and their symbols acquire deep meaning according to the emotional bonds woven over the years (TUAN, 1980; 1983; MELLO, 2003). Under these conditions, one of the tasks of the humanistic field is to reconcile, understand and decode the symbolic content of places. Since the individual is not distinct from his or her place (RELPH, 1976) and each person has a particular and personal geography (LOWENTHAL, 1982; COSGROVE, 2004), a phenomenological approach that privileges the individual in his or her lived world is necessary (FERNANDES, 2014).

A symbol has the power to suggest a whole, it transcends its condition as such and, as an integral part, it blends in with the place in which it is found. In this respect, the symbolic charge of a temple or a stadium can be much broader and more expressive than its original purpose. In fact, the cross symbolizes Christianity, the crown the monarchy (TUAN, 2012), just as the Brandenburg Gate represents one of the highest symbols of the German nation (FREITAS, 1999). Symbolism, understood as an emblem or interpretation of the meaning of a particular symbolic element (symbol), has emerged in recent decades as an extremely important concept for humanistic and cultural research - studies related to understanding the subjective dimension of place (TUAN, 1980; 2012; MELLO, 2000, 2003; FERNANDES, 2015c). According to Cosgrove (2004):

> All landscapes have symbolic meanings because they are the product of man's appropriation and transformation of the environment. Symbolism is most easily grasped in the most elaborate landscapes - the city, the park and the garden - and through the representation of landscape in painting, poetry and other arts. But it can be read in rural landscapes and even in the most apparently non-humanized landscapes of the natural environment. The latter are often powerful symbols in themselves (COSGROVE, 2004, p.108).

To understand the expressions imprinted by a culture on its landscape, we need knowledge of the language used: the symbols and their meaning in that culture. For the author, although the link between the symbol and what it represents is very tenuous, all landscapes are symbolic. By pointing out that human scenarios are loaded with symbolism, Cosgrove focuses on nature and the natural landscape as powerful symbols in themselves, starting from the assumption that any human intervention in nature involves its transformation into culture. Although this transformation is not always visible, especially to an outsider, the natural object becomes a cultural object when it is given a symbolic meaning (COSGROVE, 2004). Let's now look at Tuan's (2012) words:

> Of the many and varied reasons for moving to the suburbs, the search for a healthy environment and an informal lifestyle are among the oldest. We have repeatedly observed how the feeling for nature and rural life is encouraged by the pressures of urban life. The city environment is both seductive and irritating, beautiful and unpleasant. The rich have always been able to escape this by going off to relax in their country houses. In the Western world, the feeling for nature culminated in the romantic movement of the 18th and 19th centuries (...). The city symbolized corruption (...). The countryside symbolized life: life revealed in the fruits of the earth, in the green things that grow, in pure water and clean air, in the healthy human family (TUAN, 2012, p. 324-325).

Symbolism is not restricted to centers of wellbeing, affection, deprivation or experience, as vast, strange and distant spaces are symbols of rejection (TUAN, 1980; 2012; MELLO, 2003). Understanding symbolism as the mark of an idea - both negative and positive - of a certain symbolic element, Tuan (2012) proposes a counterpoint between the city and the countryside, suggesting that since the industrial revolution, the city - little by little - has ceased to symbolize an ideal of life, giving way to the countryside through a return to a feeling for nature. According to Tuan, by acquiring some of the values of the countryside, the suburb - understood as the frontier of metropolitan expansion - comes to represent an ideal, as it suggests a perfect lifestyle, in which the best of rural and urban life are combined without their defects (TUAN, 2012). In this sense, whether in the American context above or in the context of the Brazilian suburbs, especially in Rio de Janeiro, the metropolitan peripheries come to represent a symbol of wellbeing for their residents (CORRÊA, 2000; SOUZA, 2005; FERNANDES, 2006).

Places and symbols acquire deep meaning through the emotional bonds woven over the years. The place itself is a symbol of affection, well-wishing, satisfaction, happiness and togetherness, on the one hand, but also the scene of struggles and daily life. The symbolic nature of places establishes connections by decoding and translating their past and connecting it to the present (MELLO, 1990; 2003).

Still on the subject of the symbolic universe, let's consider the musings of geographer Doreen Massey (2008):

> And so there is "place". In the context of a world that is certainly increasingly interconnected, the notion of place (usually referred to as "local place") has acquired a totemic resonance. Its symbolic value is constantly mobilized in political arguments. For some, it is the sphere of everyday life, of real and valued practices, the geographical source of meaning, vital as a fulcrum while "the global" weaves its increasingly powerful and alienating webs. For others, "a refuge in place" represents the protection of drawbridges and the construction of walls against new invasions. Place, in this reading, is the place of negation.

In an attempt to translate the symbolic value of place, Doreen Massey (2008, p.24-25) discusses its wide range of meanings. From her perspective, place symbolizes - among other things - the sphere of everyday life, the geographical source of meaning, a vital point of support, as well as representing refuge and protection from the powerful and alienating webs of the global. Advocating a new stimulus for spatiality, the author points to nature and the natural landscape as symbolic foundations for recognizing place (MASSEY, 2008).

Delving into this universe of meanings and values, Joel Bonnemaison (2002, p.109-111) emphasizes:

> A geosymbol can be defined as a place, a route, an area which, for religious, political or cultural reasons, in the eyes of certain people and ethnic groups takes on a symbolic dimension that strengthens their identity (...). Symbols gain greater strength and prominence when they are embodied in places. Cultural space is a geosymbolic space, charged with affectivity and meaning.

The geosymbol - a concept developed by Joel Bonnemaison (2002, p.109-111) - can be understood as a place-symbol, loaded with affectivity and meaning. Among the premises defended by geographers of the humanistic horizon are those related to the symbolic content of places (COSTA, 2008) and the mosaic of symbols that reside there (MELLO, 2003, 2008). In this sense:

> The symbolic character of places reveals itself to human beings as something that precedes language and discursive reason, thus presenting certain aspects of reality, emphasizing the relationship between the symbolic and the place. These relationships are mediated by symbols, which can be a material reality that is linked to an idea, a value or a feeling. We therefore understand that symbolic mediations permeate personal attitudes towards places (COSTA, 2008, p 149).

In his discourse on the issue of cultural heritage as a set of symbols that refer to the memory of the place, Costa (2008) alludes that "the symbolism of places leads us to the concept of vernacular landscape where this character is made explicit in the set of representations, both of ancient landscapes and current ones, expressed through the knowledge and actions of man" (p.151). For the author, certain natural or cultural elements, when associated with the everyday relationships of individuals or social groups, can define a set of symbols that express the memory of the place. In these circumstances, everyday relationships and the consequent understanding of places and their symbols can turn a space into a place, once it has been affected. "In this context, place becomes more interesting as a reference for identity and at the same time acquires a symbolic value" (COSTA, 2008, p. 155).

Considering the symbolic universe of places, reconciling, understanding and decoding the symbols of Guaratiba are tasks to be undertaken in the following pages, starting with the symbolic value of its toponym (FERNANDES, 2022a).

3.4.1 Toponymy and identity

Naming places has a strong meaning, since these names establish connections between the place in the past and today. As such, toponymy reveals ownership, memory, symbolism, affection, adherence, resistance and intimacy with the place named (MELLO, 2007). In this field, Corrêa (2003, p. 176) emphasizes: "toponymy is a relevant cultural mark and expresses the effective appropriation of space by a given group, being a powerful element of identity". Under these conditions, the names of streets or neighborhoods give places a strong identity, being the result of experiences, clashes, utopias and values, among friends, relatives, strangers, acquaintances and feelings, making up a whole of introjections, strangeness, adherence and belonging (MELLO, 2000).

For Lessa (2001), toponymy is the first and most faithful record of places, a kind of baptism. According to the author, people's loyalty to their names takes on a vigorous dimension. In this line of argument, Lessa (2001, p. 58) argues:

> Names have much greater longevity than the material configurations of places. One example is the names that endure even though their original places no longer exist: they retain an unmistakable symbolic character. Praça XI, which is still evoked in Rio's samba; Castelo is the site of a hill that has already been demolished; Rua do Ouvidor, which hardly anyone knows who it was, and there are even doubts as to which ouvidor it might have been. Other places change euphoniously: the Battle of Cerro-Corá, from the Paraguayan War, gave rise to the Serra Coral slum. Praça do Asseca became Praça Seca; Willian's place became Ilha (de Guaratiba) etc.

Lessa (2001, p. 427) shows himself to be a profound connoisseur of the places in his city and of the contexts that gave rise to their names, as he approaches toponyms as a true mosaic that unites contemporary and past elements. In his quest to understand the meaning and significance of the names of different places in the city of Rio de Janeiro, the author discovered that it was "a certain Willian, owner of a farm in Guaratiba, who was responsible for the origin of the name of the place called 'Ilha'" (IMAGE 29).

IMAGE 29 - THE DOMINIONS OF A DRY ISLAND - Plaque alluding to the toponym Ilha de Guaratiba located in Praça da Ilha - the main street in the area (FERNANDES, 2019a).

Like most people who look at Ilha de Guaratiba, Lessa used the writings of the historian Rivadávia Pinto (1986) to arrive at the origins of the place's name. According to him, the place name "Ilha" originated as a corruption of the name of the Englishman Willian. Coming as part of the English escort that protected the Portuguese Royal Family on their transfer to the former Terra de Santa Cruz in 1808, William took possession of and began to live in the area that is the subject of this investigation. As the natives were not very good at pronouncing his name correctly, they started calling him "Wílha", his "Ilha de Guaratiba" and, finally, "Ilha de Guaratiba", the former owner of the local domains (LESSA, 2001). The much older toponym "Guaratiba", on the other hand, was derived from the large number of wading birds that inhabited the area - the guarás. As the word "tiba", in Tupi-Guarani, means abundance, Guaratiba, etymologically, means "abundance of guarás" (PINTO, 1986). In these terms, the toponym Ilha de Guaratiba arises from the crossing of British and indigenous words (FERNANDES, 2014d).

Although this corrupt version is widely accepted (IMAGE 30), there are other versions that try to elucidate why a place not surrounded by water is called Ilha. On old maps of the town, it was possible to see the two rivers that skirted the area, turning the place into an island, since it was surrounded by their waters. At that time, not only were the waters of the aforementioned rivers more voluminous, but they were also much more influenced by the tides. The story of "Seu Ilha" is the most credible. However, there are other stories about the place's name. The old Guaratibanos used to say that in the past the neighborhood was a kind of island surrounded by tidal channels. Many still say today that these rivers were quite large. This large volume of water increased even more when the tide rose and residents were stranded during high tide periods. The existing rivers and canals made it difficult for people to access the town. At that time, the place was isolated from time to time, depending on the flow of the rivers and tidal channels. According to this version, the place was periodically "cut off" by the floods caused by the tidal channels (FERNANDES, 2014c).

IMAGE 30 - THE DOMINIONS OF AN ISLAND CONCERNED BY THE HEIGHTS OF THE GUARATIBA GENERAL SERRA - Meme alluding to the toponym Ilha de Guaratiba which, despite its toponymy, is not an island neighborhood. In: https://www.facebook.com/GeografiasGuaratibanas. Accessed on 01/25/2024.

Guaratibanos often agree that the place name Ilha de Guaratiba comes from the name of an Englishman. Other residents, however, look to the roots of the place for another explanation. Although I don't dispute the official toponymic version based on the story of William (Seu Ilha), it's worth emphasizing that other explanations can be given for the genesis of the place's name. For example, the place name "Ilha" could be linked to the periodic flooding of part of the Baixada de Guaratiba, due to the rising waters at high tide. The bases for this theory are some historical records that assure the occurrence of regular flooding in the areas near the canal rivers of the Guaratiba Tidal Plain (PINTO, 1986; CASTRO, 2002) and archaeological (KNEIP, 1987) and biological (ARAÚJO, 1987; MENEZES, 2005) studies of the Baixada de Guaratiba. The occurrence of an ancient port on Ilha de Guaratiba would prove the existence of flowing rivers in the area. This information is confirmed by Fridman (1999), who mentions the existence of the Ilha Port, which was in full operation at the beginning of the 19th century (FERNANDES, 2021).

In recent research, based on the orality of Guaratibanos with a long tradition (FERNANDES, 2023a), I addressed issues related to the origin of the toponym "Ilha" (from Guaratiba) in its different versions. Interviewing residents of different ages and social classes, I came to the conclusion that there are at least five versions of the place's name:

1 - On old maps of the town, there were two rivers which, when the tide rose, surrounded the place like an island;

2 - The place is surrounded by the Serra Geral de Guaratiba as an island, limited by the elevations belonging to the Pedra Branca Massif;

3 - Before the construction of Avenida das Américas, the former Rio-Santos highway and now the Transoeste corridor, the tide periodically rose and washed away the area;

4 - The town was far away from everything and everyone, so that when the newspaper arrived, the news was from days gone by. The place was isolated like an island;

5 - In the past, the place belonged to an Englishman called Willian. As the old residents didn't bother to pronounce the foreigner's name correctly, they started calling him "Seu Ilha" (from Guaratiba).

Like any self-respecting scientific research, I went in search of documentary sources that could prove or disprove the rich information provided by my Guaratibano neighbors. So, when I looked at the writings of Nei Lopes (2012), based on the researcher Fania Fridman (1999), I discovered that: "The origin of the locality known as Ilha de Guaratiba goes back to the old Engenho da Ilha, a property that already existed in the 18th century". As a result, the place name "Ilha" could possibly refer to a river island, dialoguing with the first and/or third versions mentioned above. This information would therefore refute the official version of the place name, which is related to the history and geography of "Seu Ilha" (FERNANDES, 2023).

The fact that I have come to know the history and geography of the toponym "Ilha", however, does not give me the right to disregard the other versions that have emerged from the rich orality of geographical memories and the toponymic creativity of the Guaratibano people (FERNANDES, 2012).

The toponymic origin of another Guaratiba domain is also worth mentioning. This is the case of Pedra de Guaratiba. The first version of this place name refers to Pedrão, or Pedra do Salão. When returning from their journeys, the old fishermen of the town would see this natural viewpoint located on the shores of Sepetiba Bay, which they called "Pedra de Guaratiba" (IMAGE 31).

IMAGE 31 - THE DOMAINS OF A STONE EMBRACED BY THE WATERS OF SEPETIBA BAY. One of the versions linked to the origin of the place name "Pedra" (from Guaratiba) is linked to a natural viewpoint known as "Pedra do Salão" (better known as "Pedrão"). Photograph taken by the researcher on January 21, 2024.

This is how the town came to be known in reference to this rocky artifact. In the same vein, it is public knowledge that many Guaratibanos believe that the stone that gave rise to the place's name no longer exists, having been extracted from a disused quarry located nearby (PINTO, 1986).

3.4.2 **A Guaratiba of symbols**

A common human tendency is to anchor the range of attachments, belongings and feelings in our place to certain experiences and symbols from the past (MELLO, 1991; 2000; 2003; 2008; TUAN, 1982; 2012; 2013; BUTTIMER, 2015). Many thinkers share the idea that the spatial and social changes that have taken place in most communities lead their residents to think that "lost ways of life are the right ways to live" (GONDAR and DODEBEI; 2005, p. 9). An individual almost "never values what they have until they lose it" (BUTTIMER, 2015, p. 5). This nostalgic feeling is generally associated with older people (NORA, 1993; BOSI, 2003) who, with many years of experience in their respective lived universes, have many memories of their places in different temporalities. Their memories reveal geographies of yesteryear which, although they are related to past spatialities, remain alive in their memory and in their hearts. They are "memorable geographies" that they miss (FERNANDES, 2011).

Pertaining to the symbolic universe of a Guaratiba of yesteryear, the streetcars that connected its domains to other localities represent the place's past. Until the 1960s, when they stopped running, the cable cars filled Guaratibanos with pride. After all, very few places could enjoy this means of transportation at the time (IMAGE 32).

IMAGE 32 - THE ILHA TRAM . The image above focuses on the Ilha de Guaratiba streetcar line traveling along the Monteiro road in Campo Grande, then

a rural area of Rio de Janeiro. Inaugurated in 1923, the Ilha streetcar ran from Campo Grande train station to the town of Ilha, in Guaratiba. Over 20 kilometers long, the tracks of this memorable machine were managed by the municipality's Rural Transport Service (FERNANDES, 2015c).

The streetcar was an important element in the process of urban expansion in Rio de Janeiro (WEID, 1997). For Abreu (2008), the two elements responsible for the city's expansion were the trains (running on tracks from 1858) and the streetcars, running after 1868 (IMAGE 33).

IMAGE 33 - THE BONDA DA PEDRA IN 1953. The image above focuses on the streetcar that belonged to the Rural Transportation Service, a municipal transportation company for the once rural areas of Rio de Janeiro. The Pedra line left from Campo Grande train station and ended at Raul Barroso square in Pedra de Guaratiba, passing along Avenida Cesário de Melo and Estrada do Monteiro,

among other streets. In: https://www.facebook.com/GeografiasGuaratibanas. Accessed on 30/01/2024.

Trains and streetcars began to act together in terms of the connection and dance of places, if we want to use the conceptual ideas of David Seamon (1980) and Paulo Maurício Rangel (2013) in the Portuguese version published by the magazine Geograficidade and as recorded in the work of Maurício Abreu (2008) specified below.

In this sense, we can say that the streetcars connected the city's most remote neighborhoods to the stations belonging to the Dom Pedro II Railway, which was inaugurated in 1858 and extended to Santa Cruz in 1878. The railway lines that connected Ilha de Guaratiba and Pedra de Guaratiba to Campo Grande railway station were inaugurated in 1924.

With a length of approximately 20 kilometers, these routes, managed by the municipality's now-defunct rural transport service, were decommissioned in 1967, when their tracks were covered up by the asphalt of the current streets and roads. However, decades after their extinction, the Ilha and Pedra cable cars are still remembered and revered as one of the past symbols of the place. The streetcars were part of the choreography undertaken on a daily basis, transporting the working masses, students and ordinary people as they came and went. Highway fever and the establishment of the automobile industry in the country in 1957 led to the removal of streetcar tracks in Rio de Janeiro, then the state of Guanabara, during the Carlos Lacerda government (1960-1965) and the administration of Negrão de Lima (1965-1970). In this context, the vehicle known as "lotação" (IMAGE

34) - likewise - is still cited today as a symbolic element of a Guaratiba of memorable geographies (FERNANDES, 2015c; 2017; 2019).

IMAGE 34 - IN THE DAYS OF THE LOTAÇÃO . Between the 1940s and 1960s, there were few alternative roads in Rio de Janeiro, especially in the far-flung rural areas of the city (today's West Zone). Not all parts of the city were served by the few bus lines, since there were few companies involved in road transportation at the time. Streetcars, likewise, were only an alternative for a few localities and the number of vehicles was insufficient for the growing population on the outskirts of Rio. In this context, the vehicle known as the "lotação" became an alternative for many residents of the former "carioca hinterland". The "lotação" represented the alternative transportation of the time, and was not tied to any formal private or state link. The bus in the photo below ran between Campo Grande and Barra de Guaratiba, passing by Ilha de Guaratiba, in the 1960s. In: https://www.facebook.com/GeografiasGuaratibanas. Accessed on 30/01/2024.

Both places and symbols can be forged through immediate experiences. At first glance, an individual may fall in love with a particular spot. On the other hand, they may need a certain amount of time for an artifact to capture their attention, rising to the level of symbol or place (MELLO, 1991, 2003). By highlighting place as a reservoir of memories and dreams, Tuan (2013, p. 227) stresses: "the sensation of time affects the sensation of place". Let's look at an excerpt that points to the relevance of the past as a pillar of individual and collective identity and a source of meaning and symbolism:

> What can the past mean for us? People look back for various reasons, but one is common to all: the need to acquire a sense of self and identity. I am more than what is defined by the fleeting present. I am more than someone who at this moment struggles to put thought into words: I am also a writer whose book has been published, and here the book is, bound, by my side, renewing my confidence (...). To strengthen our sense of self, the past needs to be rescued and made accessible (TUAN, 2013, p.227-228).

Each person carries their place with them through experiences, familiarity, affection, belonging and other experiences. This range of feelings is woven over time and evoked, consciously or unconsciously, at all times, denoting that what we are and what we possess is the result of our history and the mosaic of experiences we have lived through in our home, understood on the most diverse scales (BUTTIMER, 2015). In this sense, "awareness of the past is an important element in the love of place" (TUAN, 2012, p.144) and history is responsible for the feeling of belonging and love for the sphere of the lived, since, over time, a person invests a significant part of their emotional life in their home and neighborhood (TUAN, 2012).

Human beings try to understand the world based on their past experience (LOWENTHAL, 1982; 1998), assimilating elements presented by formal and informal education over the course of their lives. This is, as philosophers argue, the so-called stock of knowledge (SCHUTZ, 1979). In these circumstances, "experiences in the scenarios of the past are treasures kept with great tenderness" (MELLO, 1991, p.235). In line with this premise, Harvey (1992), drawing on Rossi, cites the historical reference and the collection of the past as a source of meaning for "cultural symbols":

> The impulse to preserve the past is part of the impulse to preserve the self. Without knowing where we've been, it's hard to know where we're going. The past is the foundation of individual and collective identity; objects from the past are the source of meaning as cultural symbols. The continuity between past and present creates a sense of sequence to the random chaos and, as change is inevitable, a stable system of organized meanings allows us to deal with innovation and decay. The nostalgic impulse is an important agent of adjustment to crisis, its social emollient, reinforcing identity when confidence is weakened or threatened (HARVEY, 1992, p. 85).

Since every lived experience dates back to the past, the importance of past experiences, places and symbols in the process of building identity is unquestionable (HARVEY, 1992). This heritage links people to their lived places, which are now proclaimed and revered not only for their contemporary characteristics, but also for the history and geography that have been built by individuals and social groups on the ground they have experienced over time. As already explained, "history plays an essential role in the human sense of territoriality and place" (TUAN, 1982, p. 156). At this point, "the identity of a place is its physical characteristics, its history and how people make use of its past to promote awareness" (TUAN, 1982, p. 156).

Time passes, leaving its trail full of lived or memorialized history. This, attached to the present, corresponds to a painstaking compilation of material and immaterial symbols which, however distant they may be from the eyes, hands and other senses, are responsible for endowing our lives with meanings and values, since we are the result of our experiences lived over time (TUAN, 2013).

The domains discussed here, which were in force in memorial times in Guaratiba, make up one of the past symbolic mosaics disseminated by members of the community, including the author of this book. In addition to the symbolic elements highlighted in the previous pages, other artifacts representative of Guaratiba's domains are worth mentioning. This is the case of geographical entities such as Fazenda Mato Alto (IMAGE 35), Sítio do Pica-Pau Amarelo and Píer da Capelinha.

IMAGE 35 - THE DOMINIONS OF FAZENDA MATO ALTO - Located in the Guaratiba lowlands, Fazenda Mato Alto is a kind of remnant of the area's agrarian past. In: https://www.facebook.com/GeografiasGuaratibanas. Accessed on 21/02/2024.

Fazenda Mato Alto is a century-old property specializing in calf rearing. It is one of the largest agricultural enterprises in Rio's West Zone. Located in the neighborhood of Guaratiba, between Avenida das Américas (the current Transoeste corridor) and Estrada do Mato Alto, the aforementioned agropastoral domain is home to a private railroad, with the right to a railroad station: the E.F. Fazenda Mato Alto (IMAGE 36).

IMAGE 36 - GUARATIBANAS LOCOMOTIVES - In the domains of Fazenda Mato Alto there is a private railroad where memorable locomotives travel. In: https://www.facebook.com/GeografiasGuaratibanas. Accessed on 21/02/2024.

The Fazenda Mato Alto railroad is where the formidable locomotives used to run, which, in the middle of 2024, persist in taking us on a memorable journey back in time (IMAGE 37).

IMAGE 37 - THE TRACKS OF THE MATO ALTO RAILROAD - tracks of the private railroad of Fazenda Mato Alto. In: https://www.facebook.com/Geografias Guaratibanas. Accessed on 21/02/2024.

In addition to housing the 3.8-kilometer-long private railroad, Fazenda Mato Alto is a locomotive workshop. The work is done by the employees who also help look after the many head of cattle - which in fact guarantee the property's livelihood. The smoke trains, which travel around the farm at a maximum speed of

IMAGE 38 - A NOVELA SCENARIO - Because of its uniqueness, which harks back to a bucolic atmosphere that is unusual in Rio's urban environment, Fazenda Mato Alto has been used as a backdrop for many television productions. In the image above, some scenes from the soap opera "Amor de Mãe", which aired on Rede Globo in 2021. In: https://www.facebook.com/GeografiasGuaratibanas. Accessed on 21/02/2024.

30km/h, are used in everyday activities. The scenery of this memorable Guaratibano domain has already served as the backdrop for soap operas such as "Sinhá Moça", "Cabocla" and "Araguaia" and to this day is a veritable open-air recording studio (PICTURES 38 AND 39).

IMAGE 39 - ACTIVITIES AT FAZENDA MATO ALTO - The locomotives support the activities at Fazenda Mato Alto. In the image above we can see one of the farm's marias fumeças transporting feed to feed the property's cattle. In: https://www.facebook.com/GeografiasGuaratibanas. Accessed on 21/02/2024.

Another landmark that is highly representative of the areas covered by this book is "Sítio do Pica-Pau Amarelo" (IMAGE 40). Monteiro Lobato's fable that brought to life emblematic characters such as Saci Pererê, Visconde de Sabugosa, Cuca, Dona Benta, Emília - among others - had the Guaratiba landscape as its backdrop. The setting for this true masterpiece of Brazilian literature transformed into a television program was a beautiful site built for the recordings of the series shown on TV Globo between 1977 and 1986 (FERNANDES, 2021).

IMAGE 40 - THE DOMINIONS OF SÍTIO DO PICA-PAU AMARELO - In the image above we can see the result of the renovation that transformed the old set of the television series "Sítio do Pica-Pau Amarelo" into a museum that reproduces the stage on which the characters in Monteiro Lobato's work came to life. In: https://www.facebook.com/GeografiasGuaratibanas. Accessed on 21/02/2024.

The building used for filming the show "SITIO DO PICA PAU AMRELO" is located on the banks of the Roberto Burle Marx Road in Barra de Guaratiba. Practically all the external and internal scenes of the series were recorded in this building. The site, which had been abandoned for decades, has recently undergone a makeover with a view to transforming it into a museum that reproduces the setting in which Monteiro Lobato's characters came to life (FERNANDES, 2023).

I recently visited Sítio do Pica-Pau amarelo on a Sunday with Luigi and Nicole. My children were delighted with the simplicity and detail of the new facilities that have transformed the site once again into a magical place that takes us back to the memorable stage of this masterpiece of Brazilian literature (FERNANDES, 2022).

The Capelinha pier stretches from the Nossa Senhora do Desterro Church to the Amendoeiras Restaurant in Pedra de Guaratiba (IMAGE 41). The relief piers were built in 2004 and have since been incorporated into the Guaratiba landscape. In a recent survey, we found that these landmarks, which stretch out over the waters of Sepetiba/Guaratiba Bay, have been elevated to the status of symbols of the neighborhood by many of its residents. The pier in the image below is located at the end of Rua Barros de Alarcão, where Praia da Capela used to be, a strip of sand that no longer exists, but which has become immortalized in the geographical memory of the traditional residents of Pedra de Guaratiba.

IMAGE 41 - THE CAPELINHA PIER. In: https://www.facebook.com/ GeografiasGuaratibanas. Accessed on 21/02/2024.

Places and symbols acquire deep meaning through the emotional bonds woven over the years. The place itself is a symbol of affection, well-wishing, satisfaction, happiness and togetherness, on the one hand, but also the scene of struggles and day-to-day life. The symbolic character of places establishes connections by decoding and translating their past and connecting it to the present (MELLO, 1990; 2003). Seeking a better understanding of the aforementioned premises, I endeavored to apply them to the context of my lived universe. After all, to paraphrase Tuan (1961, P. 32), "I have no obligation to describe any area other than that for which I have a special affection or an inexplicable fascination". Guaratiba, this place/symbol, as well as exerting a strong hold over me, has the power to capture my affections (FERNANDES, 2003; 2005; 2006; 2008; 2009; 2010; 2011; 2012; 2013; 2014; 2014b; 2014c; 2014d;

2015; 2015b; 2015c; 2015d; 2016; 2017; 2017b; 2018; 2019; 2019b; 2020; 2021; 2022; 2022b; 2022c; 2023; 2023b; 2024).

3.5 **Affective domains**

Commonly, the concepts of space and place express - metaphorically and respectively - the notions of darkness and luminosity (MELLO, 2000). In this vein, far from positivist and neo-positivist impositions, there is no pre-established condition for opaque spaces, immersed in penumbra, to reach the level of place through their illumination or clarity. Paraphrasing Mello (2003), we can point out that a place can lose or gain this status depending on the darkness or brightness with which it becomes overshadowed or even illuminated over time. According to Tuan (2013, p.199), "place is any stable object that captures our attention". However, the same thinker ponders: "many places, highly significant to certain individuals and groups, have little visual notoriety" (TUAN, 2013, p. 200). However opaque or visible they may be, certain objects or places admired by one person may not be noticed by another (TUAN, 2013).

Place is, par excellence, a reservoir of memories and dreams (TUAN, 2013). In these circumstances, visual notoriety alone is not an instrument for certain spaces to become places. A deep sense of place results from a combination of historical, geographical, cultural, economic, existential, subjective, intersubjective, immaterial and visual factors and values (YÁZIGI, 2003). With regard to the line of thought undertaken in this context, visibility is not only linked to human constructions and elements of nature that give visual values to certain artifacts. However, there is no denying that these natural and cultural attributes, in many cases, represent factors in the transformation of certain localities which, as a result, can undergo qualitative or non-qualitative changes, depending on the different perspectives of those who experience them. With regard to the different viewpoints of individuals in relation to visibility, Tuan stresses that most places are not deliberate creations, as they are built to satisfy practical needs. In this context, places acquire visibility and meaning for both locals and outsiders (TUAN, 2013).

The boundaries of a place are fluid and existentially demarcated by its residents and inhabitants. On the other hand, the same perimeter is unknown to outsiders, visitors and newcomers. This niche of "ambiguities, topophilic feelings, fears and people's philosophical way of acting" (MELLO, 2000, p. 129), can forge spaces and places, respectively, through the valorization and valuation of undifferentiated places, spaces that until then had been treated with relative indifference even by their insiders (FERNANDES, 2006). In this sense, visibility, related to natural and built landscape attributes, becomes a significant element, both for insiders - who start to value the lived place more after its valorization - and for new residents who see these peculiarities as indispensable factors for the valorization of a space that for them was undifferentiated, thus making this lived universe a home or place.

Embracing this philosophy about the places of the men and women focused on in our investigation, we understand that the process of real estate development that has given visibility to residential areas such as Ilha de Guaratiba, has promoted significant changes in the existential way of life of the local residents. It is these existential metamorphoses that flourish in the midst of this urbanization process that will be discussed below, starting with the role played by valuation in the phenomenon of transforming space into place.

3. 5.1   From valuing space to valuing place

Valuation - here understood as absolute value or objective value, price or economic value attributed to a given object, asset or area - when related to the forms of housing directed by the real estate market, can become enlightening to elucidate the dynamics involving the phenomenon of place making. In this sense, visibility, understood as the visual character of a given locality, may be the main factor responsible for that area achieving a certain status related to its real estate value, as is the case with most neighborhoods with significant income strata (TUAN, 2013).

In the US context, even in the first half of the last century, Walter Firey (1945; 2013) was already looking at the network of relationships between the symbolic and sentimental charge of places (valuation) and their residential attractiveness for the economically privileged classes (valorization). In Firey's view, certain places should not only be attributed the economic variable, since they are also endowed with a kind of "sentimental mark". Among the examples cited in Walter Firey's study is the case of an upper-income residential neighborhood known as Beacon Hill, located in central Boston. In the study in question, Firey points out that the growing economic value of the

Beacon Hill neighborhood is due above all to its symbolic charge and the sentimental associations built up over its century-long history. "On Bacon Hill, a series of aesthetic, historical and familial feelings have been spatially articulated" (FIREY, 2013, p. 23).

Regarding the link between the real estate development that is taking place on Ilha de Guaratiba (FERNANDES, 2006) and its "spatial representations" (FIREY, 2013, p. 22), it is important to consider that, from the moment people with considerable purchasing power began to buy farms or houses in this area of Guaratiba, residents who didn't value the place enough began to wonder: why are these families coming to our neighborhood? From that moment on, the younger residents realized that there was something special about their lives. Ilha de Guaratiba began to have value for these people. The rise in real estate prices brought many changes to the area. Due to real estate advertisements, many new residents migrated to Ilha de Guaratiba from other areas of Rio. This increase in people and urban activities has made younger people, in particular, value the place more. Twenty years ago, younger people thought they lived at the end of the world. They didn't like Ilha de Guaratiba. Today, as they have matured, they have come to value the place more. In addition, the environmental issue is currently in vogue. This contributes to people becoming more attached to nature. In this context, the appreciation of Ilha de Guaratiba has caused concern among its residents, who now value the bucolic nature of their place more. If people from noble and traditional neighborhoods are coming to live here, it's because the place has value, isn't it? A few years ago, young people here wanted to be in the places where a considerable number of the new residents come from. In addition, the increase in urban violence has made young Guaratibanos value their neighborhood more. The appreciation that has come from other people, from other places, has certainly contributed to us - Guaratibanos - having a more affectionate relationship with our place.

There is a maxim, commonly used in common sense, according to which "people only value things after they have lost them". This axiom reveals a recurring human tendency, since the absence of an essential good, a relevant feature or artifact, or a loved or cherished person, in many cases leads us to recognize the importance of their existence or presence. In this sense, presence and absence are directly related, since the absence of a loved one, for example, can be largely responsible for their presence in our minds and hearts (LEFEBVRE, 1983). Not only absence, but also the fear of losing something essential, can trigger a series of changes in the attitude and existential way of life of certain individuals and social groups (BUTTIMER, 2015).

The question here is simple and consists of unraveling the following question: can the process of (economic) valorization that is taking place on Ilha de Guaratiba influence the way residents experience the place? Could it be that once the place was valued from an economic point of view, it became more "valued" by its residents?

That said, we can stress that the Guaratiba residents' unease, caused by fear or suspicion of the rivalry represented by the presence of new residents in their place, may have been the preamble to a new relationship with their lived universe. In the specific case of Ilha de Guaratiba, real estate development has not only led to spatial changes. For different reasons, the *addition of* new homes and residents has triggered a change in the attitude of younger Guaratibanos who, following this increase in value, have come to "value" the place they had previously treated with indifference. In this respect, we can emphasize that this new relationship between younger people can be understood as a true "existential metamorphosis", since detachment from the former residential "space" has metamorphosed into affection for their current "place". Regarding the feelings pertinent to the changes brought about by speculation, valorization and urbanization, Guaratibanos - jealous, perplexed and frightened to see their lived world being invaded by people and families from other neighbourhoods in the midst of this invasion-succession process (CORRÊA, 2000) - begin to strengthen their ties with the lived universe even more in the condition of place.

Valuation - understood in terms of subjectivity, esteem, affective value, identity (HAESBAERT, 2004), references for the construction of spatial identities and a sense of belonging (SOUZA, 2004) - denotes the symbolic and/or philosophical value attributed to a given place, and is built on lived experiences. The transition from space to place mainly involves the affective value of the spatial portion in which the individual is inserted. In many cases, however, the economic or other value attached to an artifact, street or area can be a fundamental element in building bonds and a sense of belonging. In this way, we can infer that, in specific cases, the appreciation of a given space can produce its valuation and consequent transformation into a place (FERNANDES, 2003; 2006; 2010).

Considering that the individual is not distinct from his or her place (LOWENTHAL, 1982; COSGROVE, 2004), as the supporters of humanism in geography reiterate, we also conclude that the events that take place in a certain spatial domain can influence, directly or indirectly, the lives of those who experience it. Spatial metamorphoses are channeled by human beings and can change the way they live and even the way they relate to their lived place. In these circumstances, the emotional bond between the person and the place, forged in everyday life, can be weakened or strengthened, depending on how the changes are introjected.

On this investigative scale, we realized that, before speculation and real estate development, the place was treated with a certain indifference, especially by its younger residents. In a clear ethnocentric stance, they believed they lived at the "end of the world" or "where Judas lost his boots", wishing to migrate to brighter places, where the attributes of urbanity would provide them with a dynamic life. That said, it is worth remembering that ethnocentrism takes on the form of valuing the place and the person and, on the other hand, the internalization of negative aspects leads to the depreciation of the place lived in and of one's own self-esteem (TUAN, 1980; 2013; MELLO, 2000). For them, Ilha de Guaratiba was nothing more than an undifferentiated space, since it was not valued by them (TUAN, 1983; 2013; FERNANDES, 2003; 2006; 2010). However, after the (economic) valorization that gave the place relevance, these *insiders,* noticing their lived universe being taken over by *outsiders*, began to change their minds, developing a different feeling towards it. As a result of this phenomenon, young Guaratibanos have become fond of their lived world. This appreciation, which emerged from the appreciation, triggered a change in attitude among the residents, who began to nurture feelings for their lived universe based on a new relationship (FERNANDES 2003; 2006; 2010). The indifference, contempt, disdain, rejection, disregard, disinterest, apathy and insensitivity of yesteryear are now replaced by admiration, pride, affection, sympathy, satisfaction, love and other valued feelings responsible for verbal reports that both express intimate relationships with the place and demonstrate its expression as such. However, before trying to translate the topophilic feelings of Guaratibanos - manifested from new experiences with their lived world - we will then try to capture the anguish of Guaratibanos who are fearful of a supposed expulsion - due to economic valorization - from their lived universe.

3. 5. 2   From the promotion of place to the expulsion of the lived universe

Despite the certain indifference with which Ilha de Guaratiba was treated by its residents - a phenomenon highlighted in the previous topic - and through the series of events that brought about the spatial and existential changes also dissected in this research, intimate experiences emerged in the midst of this process that culminated in the acquisition of a place par excellence. After helping to promote an undifferentiated space to the status of a place for its residents, real estate development continued to follow its course. However, after the inauguration of the Grota Funda Tunnel, the exacerbation of this process culminated in a pressing concern on the part of Guaratibanos: the fear that the changes underway will drive them out of their lived universe. But what are the residents of Ilha de Guaratiba basing their concerns on?

In Guaratiba areas such as Ilha de Guaratiba, the delimitation established in the 1970s is still in force, taking into account only the rural characteristics of the time, with a minimum plot size of 10,000 square meters (BERTA, 2012). In this context, landowners in the area were charged the Rural Land Tax (ITR) and were exempt from the Urban Land Tax (IPTU). However, since the last IBGE census - in 2010 - the population of the city of Rio de Janeiro has been considered 100% urban. Using this subterfuge, a few months later the city council changed the legislation and, likewise, began to characterize the municipal territory as entirely urban according to the 2011 master plan (complementary law 111 enacted on 01/02/2011).

According to Berta (2012), as soon as Ilha de Guaratiba became an urban area, the municipality had to work on the necessary Urban Structuring Project (PEU) in order to define the neighborhood's urban standards and ensure occupation that would not make it impossible to maintain the natural environment and its traditional production of ornamental plants. However - the PEU, which could redefine the new urban patterns of the area and regulate the form of occupation, the minimum size of plots and the maximum number of floors - has not yet been defined.

In theory, the Urban Structuring Project is a set of rules guided by policies and actions defined to guide the physical-urban development of a group of neighboring districts with similar characteristics. In practice, however, the PEU is also used to define the IPTU that will be charged. Following this redefinition of the area in question, a plot of land measuring 20,000 or 30,000 square meters that once served a family of farmers who made a living from the land, under which the inexpensive federal tax (ITR) had been levied, now faces an annual IPTU debt of 30,000 reais (BERTA, 2012).

The issues highlighted in the last three paragraphs converge to understand the apprehension of the residents of Ilha de Guaratiba who - in the midst of the spatial, administrative and tax changes brought about by the inexorable march of urbanization - express their dilemmas, anxieties, concerns and fears in the face of a possible expulsion from their lived universe. This apprehension on the part of the Guaratibanos took on a new dimension in the post-tunnel period, when the place gained visibility. In addition, real estate speculation has further boosted real estate and land values, a factor that has intensified the "march west" in recent years.

As we can see, there are many concerns among the residents of Ilha de Guaratiba in relation to current events: accelerated population growth, construction of multi-family buildings, degradation of natural environments, making rural production (of ornamental plants) unfeasible, as well as the pure and simple change from federal taxation (ITR) to municipal taxation (IPTU). This fear is justified by the fact that many residents live on relatively large areas: 5,000m², 10,000 m² or more (FERNANDES, 2015). As most of these properties are productive, residents question the fact that the IPTU can only be levied on the built-up area (building), and not on the entire property (land), a fact that has occurred in some farms, ranches and estates.

Due to the lack of specific legislation that allows for differentiated taxation for each type of property, Ilha de Guaratiba has different realities: there are properties that, because they represent possessions, pay no tax at all; there are others that, because they prove to be productive, pay only the ITR; there are those that pay the IPTU and there are also some where there is double taxation: ITR and IPTU (FERNANDES, 2010; 2017).

In the process of spatial change in Ilha de Guaratiba, real estate speculation can be considered a two-way street. By promoting real estate appreciation, this speculative practice has helped residents to "value" their former space, elevating it to the status of a place (FERNANDES, 2006). Nowadays, however, the continuity of the process of valorization - accompanied by its nuances - can contribute to Guaratibanos being "expelled" from their lived universe. This is the contradiction: the valorization that helped elevate a once undifferentiated space to the status of a place through the residents' appreciation of their experienced ground, can today be responsible for separating Guaratibanos from their place.

We can better understand the anguish, protests, regrets and fears of Guaratibanos when we look at the relationship they have built with their neighborhood over time. We can then see that this attachment to place is the result of clashes, joys, disappointments and other affective experiences - producing a niche of experiences and feelings of value.

3.5.3   **Topophilia: intimate experiences with place**

The topophilic bonds between people and their experienced land are built up over the course of their experiences living in this part of space, an experience responsible for a strong attachment and identification with the place. The propositions that "places are loved ones deserving of consideration" (MELLO, p. 50) and that in the "symbiotic relationship between men and the environment, places should be considered people and people as places" (POCOCK, 1981, p.337) are evident in our affectionate relationship with certain places. Thus, it is possible to hear from ordinary people about their affection for a neighborhood, for example, as if that place were a person deserving of their consideration and esteem for everything it has given them throughout their lives. I myself feel complete in Ilha de Guaratiba, which leads me in the direction of the humanistic premise that "place is a stretch of the earth's surface in which man completes himself" (MELLO, 1991, p.50).

In Tuan (1980), the term topophilia is defined in different ways, including "the manifestations of human love for place" (p.106), "the affective ties of human beings with the material environment" (p.107), in short, this term is always associated with the individual's feeling towards the place (p. 129). People's affiliation, attachment, affection, love, identification and other qualitative feelings towards a certain environment, however, are established through intimate experiences, in which this physical environment is elevated to the level of place (TUAN, 1883). The intense feeling of love for a place, however, is often deepened by simple events, since

> Intimate experiences, when not exalted, go unnoticed. At the time, we don't say "this is it", as we do when admiring objects of notorious or recognized beauty. It's only when we reflect that we recognize their value. At the time, we are not aware of any drama; we don't know that the seeds of a lasting feeling have just been planted. Simple events can, over time, turn into a deep feeling for the place (TUAN, 1983, p.158).

Intimate experiences, therefore, are not necessarily linked to major events, but rather to everyday events which, despite their apparent simplicity, are also responsible for people's attachment to their place. Being personal and individualized, these intimate experiences are difficult to express (TUAN, 1983), translate and decode. When revealed through spoken or written accounts, however, these experiences manifest beautiful stories of individuals' and social groups' love for their place, where it is situated by people as their own extension. In this sense, in the midst of introjections, the individual cannot distinguish themselves from their place, in which case there is no separation between subject and object (RELPH, 1976; SCHULZ, 1979; LOWENTHAL, 1982; COSGROVE, 2004).

With regard to the feeling of belonging that makes me feel at home in Guaratiba, Schutz (1979, p. 291) argues that "feeling at home is an expression of the highest degree of familiarity and intimacy" and means, among other things, customs, values, personal habits, traditions, a peculiar way of life, made up of small important and cherished elements (SCHUTZ, 1979).

3.5.4   Ethnocentrism: place as the center / navel of the world

For many people who have heard of Guaratiba, the carioca domain in relief represents nothing more than an undifferentiated peripheral area, far from everything and everyone. On the other hand, because the passion they experience doesn't share with dull, distant thoughts, there are Guaratibanos who place their place at the center of their world. Since the place, in its condition of stability and confinement, gives us a sense of an intimate and humanized home and a niche of protection and coexistence built up in everyday life, obviously, moving away from this point of support and well-wishing is a mismatch that produces disenchantment and discontent. In fact, this is the feeling I get when I need to get away from my arena of struggles and well-wishing that I explore with singular resourcefulness (TUAN, 1983, 1998; MELLO, 1991; 2000). I miss my place when I'm traveling. In this sense, his account is in line with Schutz's (1979) phenomenological postulates when he discusses some issues related to returning home.

> What, however, should "home" mean? "Home is where you leave," says the poet. "Home is the place to which a man intends to return when he is far away," says the jurist. Home is the starting point as well as the end point. It is the zero point of the coordinate system we assign to the world in order to move within it. Geographically, "home" means a certain place on the surface of the Earth. Where I happen to be is my "domicile"; where I intend to stay is my "residence"; where I come from and where I want to return to is my "home". However, home is not just the place - my house, my room, my garden, my city - but everything

> it symbolizes. The symbolic nature of the notion of "home" is emotionally evocative and difficult to describe. Home means different things to different people. It means, of course, the parental home, the mother tongue, family, love, friends; it means a cherished landscape, "songs my mother taught", food prepared in a certain way, familiar things for everyday use, customs, personal habits - in short, a peculiar way of life, made up of small, important and cherished elements (SCHUTZ, 1979, p. 290-291).

According to Gomes (2007), the first fundamental characteristic of humanism taken up by geography refers to the unavoidable anthropocentric vision, according to which man is the measure of all things. Furthermore, human beings, individually or in groups, tend to place their lived place at the center of the world. In this context, egocentrism and ethnocentrism become universal human traits (TUAN, 2012). Based on Tuan (1980), Mello (1991, p.202) describes ethnocentrism as

> a universal phenomenon of overvaluing the "center", "navel", "healthiest" or "best place in the world" and can also be understood as collective egocentrism. People from the "center" discriminate between "us" ("superior") and "them" ("of lesser value", "of inferior culture") by looking at the latter in a "blasé" way and sometimes with apathy, sarcasm or aggression.

The notion of the center is one of the most important cultural manifestations, since people tend to place the place where they live as the most important and favorable, and as the center of their world. "All ancient peoples place themselves at the center of relationships and organize what they understand by the world in this reference." In this sense, the top is a related notion. "A whole religious and cartographic symbolism derives from this centrality, organizing the conception and geographical relationship of these peoples" (MOREIRA, 2009, p. 67). According to Indian beliefs, Mount Meru stands at the center of the world. An Iranian myth believes that the sacred mountain of Elburs is located at the center of the Earth. The name of Mount Tabor in Palestine could mean "navel". Mount Garizim, in central Palestine, undoubtedly enjoyed the prestige of being a central place, as it was called the "navel of the Earth" (ELIADE, 2007). According to ancient tradition, preserved to this day in the region, Palestine, in its condition as the highest country because it was close to the summit of the cosmic mountain, would not have been covered by the flood. A rabbinical text states: "the land of Israel was not submerged by the flood". For Christians,

> Golgotha was located at the center of the world, as it was the summit of the cosmic mountain and, at the same time, the place where Adam had been created and buried. Thus, the Savior's blood was spilled on Adam's skull, buried at the foot of the Cross, serving to redeem him. The belief that Golgotha was located at the center of the world is still preserved in the folklore of Eastern Christians (ELIADE, 2007, p.24).

From the beliefs we've mentioned, we can deduce that each oriental city was located at the center of the world. For some of these peoples, the highest point of the cosmic mountain was not only the highest point on Earth, but also the navel of the world, the point at which creation began. Some traditions explain the symbolism of the center in terms taken from embryology, according to which "the Divine Being created the world as an embryo. In the same way that the embryo was formed from the navel, God began to create the world from the navel onwards, and from there spread out in different directions" (ELIADE, 2007, p. 25). In the meantime, the world would have been created from Zion and the universe conceived from a central point. The creation of man also took place at a central point, at the center of the world (ELIADE, 2007).

According to Mesopotamian tradition, man was formed in the "navel of the Earth". Paradise, where Adam was created from clay, is located at the center of the cosmos. Paradise was the navel of the Earth and, according to a Syrian tradition, it was established on a mountain higher than all the others. Adam was created in the center of the Earth (ELIADE, 2007). Based on the premise that creation was derived from a center, we can also assume that any place founded is built in the center of the world of the individuals who established it as their home, shelter, refuge and dwelling place.

The chimera related to this type of centrality is, according to Tuan (1980; 2012), necessary for the maintenance of culture. For Tuan, when raw reality shatters the illusion that one's place is superior, culture itself may decline. Even knowing that they are not at the center of things in the literal sense, it is necessary for something of this faith to be present in small communities in order for them to thrive (TUAN, 1980). The progress that comes from the human belief that their lived universe is at the center of the world, however, would not necessarily be tied to numbers, quantities and other objective values (valuation). Centrality, in this case, would be associated with a series of subjective values (valuation) based on the subjectivity or intersubjectivity of individuals and social groups. Since subjective value is a relationship between the subject who values and the object valued, attributing value to a place means not being indifferent to it. Non-indifference is the main characteristic of value, since we are not insensitive to a particular place that captures and radiates our affection and sympathy, because we are always influenced by experiences in the place we live. Valuing is a fundamentally human experience that lies at the heart of every choice in life (ARANHA and MARTINS, 1992). Among these choices is the choice of the spatial portion that distinguishes itself from all the others, not necessarily because of its form-content, but because of the experiences, the feeling of belonging, the affection and the other qualitative values that its experiencers nurture for it, a valuation that will focus this once undifferentiated space in the navel of the world, as the center of the universe.

In approaches related to humanism in geography, the concept of place, because it has no defined scale, becomes too diffuse since it can either designate a place or encompass the planet (TUAN, 1983). However, the same Tuan who takes up the maxim that defines Geography as "the study of the Earth as the home of people" (TUAN, 1991, p. 89), also points out that "topophilia sounds false when it is manifested by an extensive territory" (TUAN, 1980, p. 116). However, patriotic love, significant in its dimension, can contradict his elucubrations. For the aforementioned geographer, topophilic feelings require "a compact size, reduced to the biological needs of the human being and the limited capacities of the senses". Furthermore, a person can identify more easily with an area if it appears to be a natural unit, small enough to be known personally (TUAN, 1980, p. 116-117). In this field, place is confused with the local sphere, and is also the locus of everyday life "responsible for human passions through communicative action and various manifestations of spontaneity and creativity" (SANTOS, 2002, p. 322).

Ilha de Guaratiba is one of those small communities, proclaimed by residents like me, a neighborhood that daily life has transformed into a center full of values, an inseparable whole made up of people, friends, acquaintances, relatives, territorial base, evocations and other references, such as "songs that my mother taught me" (SCHUTZ, 1979) that allow those who experience it the pleasant sensation of feeling at home or even emersed in the center-umbiglo-heart of the world (TUAN, 1980, 1983, 1998; MELLO, 1991; 2000).

3.5.5   **Guaratiba: a place**

A "place is any stable object that captures our attention" (TUAN, 2013, p. 199). Its boundaries are existentially demarcated for its residents and experiencers. There are propositions that "places are loved ones worthy of consideration" and that in the "symbiotic relationship between men and the environment, places should be considered people and people as places" (POCOCK, 1981, p.337).

What is a place? What characteristics can be listed to elevate a locality to the status of a place? Is Guaratiba a place? What makes this simple locality a place? These last questions cannot be answered on the basis of theories, since the answers are linked to the lived experiences of individuals and social groups in their lived universe. Only they can talk about their experiences, feelings and meanings built up over the course of time on their experienced ground. Only the intensity of these experiences, as well as their interpretation, can define whether or not the place represents a place for its residents.

Based on my 51 years of lived experience, I can say that Guaratiba is a place. As a representative of the community, I am a resident who is passionate about my lived universe. Among the reasons and/or characteristics that guide Guaratiba's "locality" (RELPH, 2012), I would like to mention: the importance of the place for my formation as an individual and a person; my passion for the place as if it were a person; my deep affection for the place, built up and strengthened over time; my sense of belonging and rootedness to the place; the nostalgia I feel for the place when I'm far from the center of my world and my deep affection for Guaratiba's domains.

For me, Guaratiba is "the place par excellence". The cradle of my history and geography, the place has always welcomed me throughout my years of existence: the place where my parents were born and where I was born and raised. The place where I studied and worked. The place where I now teach in the schools where I learned to love geography. The place where Nicole and Luigi, my children, took their first steps. The place that formed me and made me who I am today. A place where I had my most profound experiences. As far as I'm concerned, there's no place like Guaratiba.

FINAL CONSIDERATIONS

Domains, limits, borders and obstacles are issues that are often focused on in geographical studies (MELLO, 2001). As we have argued in this book, the boundaries of Guaratiba concern different spaces and/or places. These, in turn, have characteristics that are similar - especially those related to their common history, what I call the 'original Guaratiba' - and singularities that differentiate the 'distinct Guaratibas' that have been consolidated in the course of their respective histories and geographies (FERNANDES, 2020). Thus, it is not appropriate to speak of Guaratiba as if this Carioca domain represented a homogeneous unit. In fact, dissimilar "Guaratibas" coexist: Guaratiba, Barra de Guaratiba, Pedra de Guaratiba and Ilha de Guaratiba.

In this vein, when issues related to real estate development and the march of urbanization are addressed, Ilha de Guaratiba is the focus of analysis. In the same way, when Burle Marx's relationship with the place is highlighted, in addition to Ilha de Guaratiba (the largest producer of ornamental plants in the state), we can include Barra de Guaratiba - the location of Sítio Roberto Burle Marx and the place where the landscaper lived for decades. Similarly, it can be said of the other Guaratiba neighborhoods: Guaratiba - the municipality's largest neighborhood - although marked by population density, still has large agricultural properties such as Fazenda Mato Alto. Pedra de Guaratiba, on the other hand, is a neighborhood strongly influenced by Sepetiba Bay, which should be called Guaratiba Bay, since it is formed by the junction between Restinga da Marambaia (which begins to form in Barra de Guaratiba) and the lands of Pedra de Guaratiba, a neighborhood surrounded by the waters of the aforementioned bay.

What unites these neighborhoods? What differentiates them? What are the characteristics that distinguish the different localities in Guaratiba? What are the symbols and/or symbolic geographies of these spaces and/or places as understood by their residents? These were the questions we sought to answer in the course of this approach, which focused on the geographies of Guaratiba.

The records of Guaratiba, it is worth repeating, date back to 1579, the year the sesmaria of the same name was inaugurated. Its current domains coincide with the boundaries established by decree 3158 of 1881 - the document that established the R. A. de Guaratiba - which is made up of three official neighborhoods (Guaratiba, Barra de Guaratiba and Pedra de Guaratiba).

In addition to the three neighborhoods mentioned above, in the interior of Guaratiba (the city's largest neighborhood), there is another neighborhood elected by its residents, which the city does not recognize as such, whose boundaries are fluid and existentially demarcated: Ilha de Guaratiba.

From this perspective, we can say that Guaratiba's domains are not just about the boundaries imposed by decree. In fact, throughout its history and geography, Guaratibanos with a long tradition have forged a memorable geography through their intense relationship with the place, based on domains of well-wishing.

Thus, we can say that the fishermen of Pedra de Guaratiba are very fond of their daily relationship with the Guaratiba/Sepetiba bay. In this context, centuries-old artifacts such as the fish market and the headquarters of the fishermen's colony are elevated to the status of place or home, since they are experienced by workers who depend on them in their daily work.

The traditional residents of Ilha de Guaratiba are also very fond of the land. The great agricultural production of yesteryear made Guaratiba one of the most prosperous parishes in old Rio in the 18th and 19th centuries. The area's agricultural aptitude prevailed until the 1980s, when food production began to decline. The end of this activity, however, did not dampen the impetus of the traditional farmers who, under the influence of Roberto Burle Marx, replaced the gardens of yesteryear with the gardens of today. Thus, through the Guaratibanos' love of the land, Ilha de Guaratiba migrated from traditional agriculture to landscaping, becoming one of Rio's gardens.

Guaratiba, like other points in the city, goes beyond its boundaries through a pulsating geography woven by its residents, having an extraordinary reach, especially when you remember that it attracts people from different places. This influx of people is both temporary (linked to visits to monuments, the Roberto Burle Marx site, restaurants, gardens...) and permanent, linked to the migration of residents from other perimeters to the domains of Guaratiba, the last frontier of the city of Rio de Janeiro.

In the year of its 445th anniversary (05/03/2024), Guaratiba continues its urbanization march, without - however - losing its charms.

REFERENCES

ABBAGNANO, Nicola. Dictionary of Philosophy. São Paulo: Martins Fontes, 2007. 1210 p.

ABREU, Maurício de Almeida. The City, the Mountain and the Forest. In: ABREU, Maurício de Almeida. Nature and Society in Rio de Janeiro. Rio de Janeiro: Biblioteca Carioca, 1992. p. 55-103.

______. Urban evolution of Rio de Janeiro. Rio de Janeiro: IPP, 2008. 155 p.

ARANHA, Maria Lúcia de Arruda. History of Education. São Paulo: Moderna, 1996. 255 p.

ARANHA, Maria Lúcia de Arruda; MARTINS, Maria Helena Pires. Philosophy themes. São Paulo: Editora Moderna, 1992. 232 p.

ARAÚJO, Dorothy Sue Dunn de. The vegetation of the Guaratiba-Sepetiba lowlands. In: KNEIP, Maria Lina et al. Prehistoric Collectors and Fishermen of Guaratiba-Rio de Janeiro. Rio de Janeiro: UFRJ; Niterói: EDUFF, 1987. p. 47-72.

ASSIS, Lenilton Francisco de. Turismo de Segunda Residência: a Expressão Espacial do Fenômeno e as Possibilidades de Análise Geográfica. Revista Território, Rio de Janeiro: Sep/Oct, p. 107-122, 2003.

ATLAS of nature conservation units in the state of Rio de Janeiro, 1990. Irregular pagination.

BERTA, Ruben. Guaratiba: Urban Structuring Plan foresees 4-storey buildings. O Globo newspaper, Rio de Janeiro, April 28, 2012.

1BONNEMAISON, Joel. Journey around the Territory. In: CORRÊA, Roberto Lobato; ROSENDAHL, Zeny (Org). Cultural Geography: a Century (3). Rio de Janeiro: EdUERJ, 2002. p. 83-131.

BOSI, Ecléa. Memória & sociedade - Lembrança de velhos. Companhia das Letras: São Paulo, 2003. 488 p.

BUTTERFIELD, Roger. "Henry Ford, the Wayside Inn, and the Problem of 'History is Bunk'". Massachusetts Historical Society Preccedings, Massachusetts, v. 77, p. 53-66,1965.

BUTTIMER, Anne. Apprehending the Dynamism of the Lived World. In: CHRISTOFOLETTI, Antônio (Org). Perspectives on Geography. São Paulo: DIFEL, 1982.
p. 165-193.

______. Home, Horizon of Reach and the Sense of Place. Revista Geograficidade, Niterói, v.5, n.1, p. 4-19, summer 2015.

CALS, Soraia. Roberto Burle Marx: a Photobiography. Rio de Janeiro: Bolsa de Arte, 1995. Irregular pagination.

CARVALHO, Ronaldo Cerqueira de. Rio de Janeiro: A city connected by tunnels - a panorama up to the end of the 1960s. 2002. Monograph (Specialization in Geography) - Postgraduate Course in Geography, Rio de Janeiro State University, 2002. Irregular pagination.

CARVALHO, Ronaldo Cerqueira de. Rio de Janeiro - a city connected by tunnels. CITY HALL OF RIO DE JANEIRO. Secretaria Municipal de Urbanismo: Instituto Municipal de Urbanismo Pereira Passos, Rio de Janeiro, 2004. 57 p.

CASTRO, Augusto César de. Guaratiba: Yesterday and Today. Course Conclusion Paper (History Degree) - History Degree Course, Fundação Educacional Unificada Campograndense (FEUC), Rio de Janeiro, 2002.

CORRÊA, Magalhães. O Sertão Carioca. In: Revista do Instituto Histórico e Geográfico Brasileiro - volume 167. Rio de Janeiro: Imprensa Nacional, 1936. 478 p.

CORRÊA, Roberto Lobato. The Environment and the Metropolis. In: ABREU, Maurício de Almeida. Nature and Society in Rio de Janeiro. Rio de Janeiro: Biblioteca Carioca, 1992. p. 27-36.

______. Urban Space. 4.ed. São Paulo: Ática, 2000. 94 p.

COSGROVE, Denis. Geography Is Everywhere: Culture and Symbolism in Human Landscapes. In: CORRÊA, Roberto Lobato; ROSENDAHL, Zeny (Org). Landscape, Time and Culture. Rio de Janeiro: EdUERJ, 2004. p. 92-123.

COSTA, Otávio. Memory and Landscape: in search of the symbolic of places. Revista Espaço e Cultura, Rio de Janeiro, Commemorative edition 1993-2008, p. 149-156, 2008.

DREW, David. Human-Environment Interactive Processes. Rio de Janeiro: Bertrand Brasil, 2002. 224 p.

ELIADE, Mircea. Myth of the eternal return. São Paulo: Mercuryo, 2007. 175 p.

FERNANDES, Marcio Luis. Guaratiba Island: From Space to Place. 2003. 44f. Monograph (Graduation in Geography) - Moacyr Sreder Bastos University Center, Rio de Janeiro, 2003.

______. Guaratiba Island in its natural attributes. Rio de Janeiro, 2005. 41f. Available at: <http://www.webartigos.com/artigos/ilha-de-guaratiba-em-seus-atributos-naturais/136345/>. Accessed on: March 29, 2024.

FERNANDES, Marcio Luis. The valorization of "Space" producing the valorization of "Place:" The case of Ilha de Guaratiba - R.J. 2006. 56 f. Monograph (Specialization in Geography) - Institute of Geography, Rio de Janeiro State University, Rio de Janeiro, 2006.

______. Guaratiba Island and its landscapes. Rio de Janeiro, 2008. 32 f. Available at: <http://www.webartigos.com/artigos/ilha-de-guaratiba-e-suas-paisagens/ 136627/>. Accessed on March 29, 2024.

______. For a Necessary Change in Values: a proposal for the production of (urban) space that favors use and not exchange. In: SIMPÓSIO NACIONAL O RURAL E O URBANO NO BRASIL, 2., 2009, Rio de Janeiro. Proceedings...Rio de Janeiro: UERJ, 2009.

______. Decoding the past and present geographies of Ilha de Guaratiba. 2010. 99f. Dissertation (Master's in Geography) - Institute of Geography, Rio de Janeiro State University, Rio de Janeiro, 2010.

______. Descortinando a marcha urbanizadora em Ilha de Guaratiba a partir das experiências vividas pelos seus moradores. Rio de Janeiro, 2011. 25 f. Available at: <http://www.webartigos.om/artigos /descortinando-a-marcha-urbanizadora -em-ilha-de-guaratiba-partir-das-experiencias-vidas-por-seusmoradores/136505/>. Accessed on March 29, 2024.

______. The Identity Character of Toponymy. In: CONGRESSO INTERNACIONAL DO NÚCLEO DE ESTUDO DAS AMÉRICAS, 3., 2012. Rio de Janeiro. Proceedings... Rio de Janeiro: UERJ, 2012.

______. Globalization and urbanization of the world. Rio de Janeiro, 2013. Available at: <http://www.webartigos.com/artigos/globalizacao -e-urbanizacao-do-mundo/136480/>. Accessed on March 29, 2024.

FERNANDES, Marcio Luis. Another horizon in search of the humanization of geography. Revista Geograficidade, Niterói, v.4, n.1, p. 78-87, verão 2014.

______. Race to the West Zone. O Dia newspaper (Opinion column): Rio de Janeiro, Nov. 26, 2014 b.

______. Unveiling the symbolic universe of a place. Revista Perspectiva Geográfica, Cascavel, v.9, n.11, 2014 c.

______. Discovering Rio - Guaratiba Island. Interview given to Band News FM Radio: 2014 d. available at: https://m.youtube.com/watch?v=PXUpRFr7Dik&feature=share. Accessed on March 29, 2024.

______. The role of secondary residences in the urbanization process on the outskirts of Rio de Janeiro: the case of Ilha de Guaratiba (article). Revista Expressões Geográficas, Florianópolis, n.10, 2015 .

______. The valorization of space producing the valorization of place (monograph abstract). Revista Expressões Geográficas, Florianópolis, n.10, 2015 b.

______. Guaratiba Island: a place unveiled by its residents and flowing into Olympic Rio. PhD thesis in geography: UERJ, 2015 c.

______. From Willian to the Grota Funda Tunnel: the march west. Paper (expanded abstract) accepted and presented at the Rio 450 Seminar - Territories and Landscapes in transformation: UFRJ, 2015 d. available at: https://www.webartigos.com/%E2%80%A6/de-willian-ao-tunel-da-%E2%80%A6/139456. Accessed on March 29, 2024.

______. Place in its multidimensionality. GeoUERJ Magazine: Rio de Janeiro, n. 28, 2016.

FERNANDES, Marcio Luis. A Place in Rio. Rio de Janeiro: New Academic Editions, 2017.

______. Representations of urban space. Revista Geografias: Belo Horizonte, v.14, n.1, 2017 b.

______. The emergence of valuation from the process of valorization: a debate on some representations of urban space. Revista Uniabeu: Nova Iguaçú, v.11, n.27, 2018.

______. An island of symbols. Rio de Janeiro: New Academic Editions, 2019.

______. Guaratiba: 440 years of memorable geographies. O Dia newspaper (Opinion column): Rio de Janeiro, March 29, 2019 b.

______. Memorable geographies. Rio de Janeiro: Novas Edições Acadêmicas, 2020. 191 p.

______. Geographies of Guaratibanas. Rio de Janeiro: Novas Edições Acadêmicas, 2021. 109 p.

______. Place and symbolism. Rio de Janeiro: New Academic Editions, 2022.

______. Guaratiba and its domains. In: RIBEIRO, Miguel Angelo; NUNES, Nathan da Silva. Perspectivas Geo-escalares sobre o território fluminense. Rio de Janeiro: Appris, v.1, 2022 b.

______. For a more human geography. Rio de Janeiro: New Academic Editions, 2022.

______. Place and Memory. Rio de Janeiro: New Academic Editions, 2023.

______. On the memory of places. Rio de Janeiro: 2023 b. Available at: https://meuartigo.brasilescola.uol.com.br/geografia/ sobre-a-memoria-dos-lugares.htm. Accessed on March 29, 2024.

FERNANDES, Marcio Luis. The geographical concept of neighborhood. Rio de Janeiro: 2024. Available at: https://meuartigo.brasilescola .uol.com.br/geografia/o-conceito-geografico-de-bairro.htm. Accessed on March 29, 2024.

FIREY, Walter. Feelings and symbolism as ecological variables. In: ROSENDAHL, Zeny; CORRÊA, Roberto Lobato. Cultural geography: an anthology, volume II. Rio de Janeiro: EdUERJ; 2013. p. 21-34.

FREITAS, Inês Aguiar de; PERES, Waldir Rugero; RAHY, Ione Salomão. Hitler's Window. GeoUERJ - Journal of the Geography Department. Rio de Janeiro n. 6, p. 29-36, 1999.

FRIDMAN, Fania. Donos do Rio em nome do rei: uma história fundiária da cidade do Rio de Janeiro. Rio de Janeiro: Jorge Zahar Editor (Editora Garamond), 1999. 304 p.

GALVÃO, Maria do Carmo Corrêa. Focus on the environmental issue in Rio de Janeiro. In: ABREU, Maurício de Almeida. Nature and Society in Rio de Janeiro. Rio de Janeiro: Biblioteca Carioca, 1992. p. 13-26.

GOMES, Paulo César da Costa. Geography and Modernity. Rio de Janeiro: Bertrand Brasil, 2007. 366 p.

GONDAR, Jô; DODEBEI, Vera (Orgs). What is social memory? Rio de Janeiro: Contra Capa Livraria, 2011. 160 p.

HALBWACHS, Maurice. Collective Memory. Presses Universitaires de France: Paris, 1968.

HALLEY, Bruno Maia. The Neighborhood and the plots of place. Revista Geograficidade, Niterói, v.4, n.1, p. 43 - 57, Verão 2014.

HARVEY, David. The Post-Modern Condition. São Paulo: Edições Loyola, 1992.

HIGHET, Gilbert. The Classical Tradition: Greek and Roman Influences on Western Literature. Oxford: Clarendon, 1949.

HOLZER, Werther. Eric Dardel's Phenomenological Geography. In: CORRÊA, Roberto Lobato; ROSENDAHL, Zeny (Org). Matrices of cultural geography. Rio de Janeiro: EdUERJ, 2001. p. 103-122.

______. Humanist Geography: a review. Revista Espaço e Cultura, Rio de Janeiro, Commemorative edition 1993-2008, p. 137-147, 2008.

JANOT, Luiz Fernando. On the road to Guaratiba. O GLOBO newspaper (opinion column). Edition published on October 26, 2013.

LEFEBVRE, Henri. Presence and Absence. Contribution to the theory of representations. Mexico: FCE, 1983. Irregular pagination.

LESSA, Carlos. O Rio de Todos os Brasis: Uma Reflexão em Busca de Auto-Estima. 2. ed. Rio de Janeiro: Record, 2001. 478 p.

LOWENTHAL, David. Geography, Experience and Imagination: Towards a Geographical Epistemology. In: CHRITOFOLETTI, Antônio. Perspectives on Geography. São Paulo: DIFEL, 1982. p. 103-141.

______. The Past Is a Foreign Country. Cambridge: Cambridge University Press, 1985.

______. How we know the past. Projeto História - Revista do programa de estudos pós-graduados de história, São Paulo, v. 17, p. 63-201, 1998.

MASCARENHAS, Gilmar. The Place of the Free Fair in the Big Capitalist City: Conflict, Change and Persistence (Rio de Janeiro:

1964-1989). 1991. 220 f. Dissertation (Master's in Geography) - Federal University of Rio de Janeiro, Rio de Janeiro, 1991.

MASSEY, Doren. Through Space: A New Politics of Spatiality. Rio de Janeiro. Bertrand Brasil, 2008. 312 p.

MELLO, João Baptista Ferreira de. Humanistic Geography: The Perspective of Lived Experience and a Radical Critique of Positivism. Revista Brasileira de Geografia, Rio de Janeiro, p. 91-115, 1990.

______. The Rio de Janeiro of Brazilian Popular Music Composers - 1928/1991 - an introduction to humanistic geography. 1991. 300 f. Dissertation (Master's in Geography) - Federal University of Rio de Janeiro. Rio de Janeiro, 1991.

______. The Humanization of Nature - an odyssey for the (re)conquest of paradise. In: SILVA, S. T.; VIANA, O. M. Geografia e Questão Ambiental. Rio de Janeiro: IBGE, 31-40, 1993.

______. In Defense of Individuals in Geographical Studies. In: NATIONAL MEETING ON THE HISTORY OF GEOGRAPHICAL THINKING 1. 1999, Rio Claro. Annals... Rio Claro: UNESP, 1999. p. 113-118.

______. From Spaces of Darkness to Places of Extreme Luminosity - The Universe of Star Marlene as a Document for the Construction of Geographical Concepts. 2000. Thesis (Doctorate in Geography) - Federal University of Rio de Janeiro, Rio de Janeiro, 2000.

______. Unveiling and (Re)thinking Spatial Categories Based on the Work of Yi-Fu Tuan. In: CORRÊA, Roberto Lobato; ROSENDAHL, Zeny (Org). Matrices of cultural geography. Rio de Janeiro: EdUERJ, 2001. p. 87-101.

MELLO, João Baptista Ferreira de. The Restoration of Places of the Past. Revista GeoUERJ, Rio de Janeiro, n.12. p. 63-69, 2002.

______. On the Pulse of the Marvelous City of São Sebastião do Rio de Janeiro. In: SOCIEDAD LATINOAMERICANA DE ESTUDIOS SOBRE AMERICA LATINA Y EL CARIBE - SOLAR, 9, 2004, Rio de Janeiro. Proceedings... Rio de Janeiro: [S.n.], 2004.

______. Values in Geography and the Dynamism of the Lived World in the Work of Anne Buttimer. Space and Culture. Rio de Janeiro, v. 19-20, p. 33-40, 2005.

______. The Drums and Arrows of São Sebastião do Rio de Janeiro. Revista Imaginário e Arte, São Paulo, n.15 p. 37-67, 2007.

MENDILOW, Adam Abraham. Time and the novel. London: John Spencer/Badger, 1960.

MENEZES, Luiz Fernando et al (Org). Natural History of Marambaia. Rio de Janeiro: EDUR, 2005.

MONBEIG, Pierre. New studies in Brazilian human geography. São Paulo: European Book Diffusion, 1957.

MORAES, Antônio Carlos Robert de. Geografia: pequena história crítica. São Paulo: Annablume, 2007. 152 p.

MOREIRA, Ruy. Brazilian geographical thought: the matrices of renewal. São Paulo: Contexto, 2009. 172 p.

NORA, Pierre. Between memory and history: the problem of places. São Paulo: PUC-SP. 1993. p. 7-28. (Projeto História, n. 10).

PARK, Robert Ezra. The City: suggestions for the investigation of human behavior in the urban environment. In: VELHO, Otávio Guilherme. The Urban Phenomenon. Rio de Janeiro: Zahar, 1976. p. 26-67.
PINTO, Rivadávia. Guaratiba: A Pride of 407 Years. Razão: the positive newspaper. Rio de Janeiro, not paginated, November. 1986.

POCOCK, Douglas. Place and the novelist. Transactions of the Institute of British Geographers: New Series, v. 6, p. 87-98, 1981.

REDONDO, Andrea Albuquerque. The city grows towards Guaratiba. In: http:// urbe carioca.blogspot.com.br. Published on September 25, 2012.

RELPH, Edward. Place and Placelessness. London: Pion, 1976. 156 p.

______. Reflections on the emergence, aspects and essences of place. In: MARANDOLA JR, Eduardo et al. What is the space of place? Geography, epistemology, phenomenology. São Paulo: Perspectiva, 2012. 307 p.

RIBEIRO, Luiz César de Queiroz. Dos cortiços aos condomínios fechados: as formas de produção da moradia na cidade do Rio de Janeiro. Rio de Janeiro: Civilização Brasileira: IPPUR, UFRJ: FASE, 1997. 352 p.

RIBEIRO, Miguel Ângelo; COELHO, Maria do Socorro Alves. The importance of the second-home phenomenon and its implications for leisure-summer activities: the example of the state of Rio de Janeiro. In: ENCONTRO NACIONAL DA ANPEGE, 9., 2007, Niterói. Proceedings... Niterói: ANPEGE, 2007. 1 CD-ROM.

RUA, João. Urbanities and New Ruralities in the State of Rio de Janeiro: Some theoretical considerations. In: MARAFON, Gláucio José; RIBEIRO, Marta Foeppel (Org). Studies in Fluminense Geography. Rio de Janeiro: Infobook, 2002. p. 27-42.

SÁ, Fátima. Burle Marx Is Not Dead. O Globo Magazine. Year 5. n.°227. Rio de Janeiro: November 30, 2008. Irregular pagination.

SANTOS, Milton. Space and Method. São Paulo: Nobel, 1992. 88 p.

______. Metamorphoses of Inhabited Space. 5. ed. São Paulo: Hucitec, 1997. 117 p.

______. The Nature of Space: Technique and Time, Reason and Emotion. São Paulo: EDUSP, 2002. 384 p.

SANTOS, Noronha. The parishes of old Rio. Rio de Janeiro: Edições O Cruzeiro, 1965.

SCHUTZ, Alfred. Phenomenology and social relations. Rio de Janeiro: Zahar, 1979. 319 p.

SEAMON, DAVID. Body-subject, time-space routines, and place-ballets. In: BUTTIMER, Anne and SEAMON, David. The Human Experience of Space and Place. New York: St. Martin's Press, 1980. p. 148-165.

SOARES, Maria Therezinha de Segadas Soares. Aspectos da geografia carioca: Rio de Janeiro: Conselho Nacional de Geografia, 1962.

______. The geographical concept of neighborhood and its exemplification in the city of Rio de Janeiro. In: BERNANDES, Lysia; SOARES, Maria Therezinha de Segadas. Rio de Janeiro - City and Region. Rio de Janeiro: Biblioteca Carioca, 1990. p. 105-120.

SOUZA, Marcelo Lopes de. O bairro contemporâneo: ensaio de abordagem política. Revista Brasileira de Geografia, Rio de Janeiro, p. 139-172, 1989.

______. Changing the City: A Critical Introduction to Urban Planning and Management. 3. ed. Rio de Janeiro: Bertrand Brasil, 2004. 556 p.

______. ABC of Urban Development. 2.ed. Rio de Janeiro: Bertrand Brasil, 2005. 190 p

TUAN, Yu-Fu. Topophilia or sudden encounter with landscape. Landscape, v.11, n. 1, p. 29-32, 1961.

______. Humanistic Geography. In: CHRITOFOLETTI, Antônio. Perspectives on Geography. São Paulo: DIFEL, 1982. p. 143-164.

______. Space and Place: The Perspective of Experience. São Paulo: DIFEL, 1983. 250 p.

______. The good life. Madison: The University of Wisconsin Press, 1986. 191 p.

______. A view of geography. Geographical Review. New York, v. 81 n. 1: p. 99-106, 1991.

______. Escapism. Baltimore: The Johns Hopkins University Press, 1998. 245 p.

________. Landscapes of Fear. São Paulo: UNESP, 2005. 373 p.

________. Topophilia: A Study of Perception, Attitudes and Values of the Environment. Londrina: Eduel, 2012. 344 p.

________. Space and Place: The Perspective of Experience: Londrina, PR: EDUEL, 2013. 248 p.

WEID, Elisabeth von der. The streetcar as an element of urban expansion in Rio de Janeiro. Rio de Janeiro: Casa de Rui Barbosa Foundation, 1997. 30 p.

WIRTH, Louis. Urbanism as a way of life. In: VELHO, Otávio Guilherme. The Urban Phenomenon. Rio de Janeiro: Zahar, 1976. p. 90-113.

YÁZIGI, Eduardo. Urban Environmental Heritage: remaking a concept for urban planning. In: CARLOS, Ana Fani Alessandri; LEMOS, Amália Inês Geraiges (eds). Urban Dilemmas: New Approaches to the City. São Paulo: Contexto, 2003. p. 253-265.

https://freewalkertours.com/pt-br/historia-rio-de-janeiro/.

https://oglobo.globo.com/rio/lenda-em-guaratiba-guara-pode-voltar-para-casa-23275075.

https://www.researchgate.net/figure/Mapa-da-divisao-administrativas-de-freguesias-do-Rio-de-Janeiro-em-1900-32-Evolucaofig 3295861217.

https://vejario.abril.com.br/cidade/pedra-do-telegrafo-melhores-passeiosmundo/?Fbclid=IwAR0lfzxLhHEUIenkAHPnFfklTxFK5dzg7 TXAH5a EEYAfw1xE Q3z-XNdX5CQ.

https://www.facebook.com/Geografias Guaratibanas.

https://bafafa.com.br/mais-coisas/sustentabilidade/reserva-biologica-de-guaratiba-ecossistema-intocado-e-protegido-pela-unesco.

ABOUT THE AUTHOR

Marcio Luis Fernandes is a teacher, geographer, researcher and writer. He was born on January 25, 1973 in Ilha de Guaratiba, a neighborhood in the city of Rio de Janeiro where he still lives today.

The son of housewife Vandi Bastos de Moraes and gravedigger Manoel Fernandes, he attended elementary school at the Narcisa Amália Municipal School (1980-1989) and secondary school at the Freire Alemão State College (1990-1992).

In 2001, she began her degree in geography at UNIMSB. In 2005, she went to the State University of Rio de Janeiro - UERJ - where she specialized in territorial policies in the state of Rio de Janeiro (2005-2006), a master's degree (2008-2010) and a doctorate (2012-2015) in geography.

In September 2023, the aforementioned researcher began his post-doctoral internship at the State University of Rio de Janeiro - UERJ - under the supervision of Professor Miguel Ângelo Ribeiro. This book, in fact, represents a substantial part of the activities carried out during his post-doctorate in geography, due to be completed in September of this year.

The father of Nicole Bazzani Fernandes and Luigi Bazzani Fernandes, Marcio Luis Fernandes is currently a geography teacher in the public schools of the state and city of Rio de Janeiro.

OTHER BOOKS BY THE AUTHOR PUBLISHED BY THE SAME PUBLISHER:

I want morebooks!

Buy your books fast and straightforward online - at one of world's fastest growing online book stores! Environmentally sound due to Print-on-Demand technologies.

Buy your books online at
www.morebooks.shop

Kaufen Sie Ihre Bücher schnell und unkompliziert online – auf einer der am schnellsten wachsenden Buchhandelsplattformen weltweit! Dank Print-On-Demand umwelt- und ressourcenschonend produzi ert.

Bücher schneller online kaufen
www.morebooks.shop

info@omniscriptum.com
www.omniscriptum.com

MIX
Papier aus verantwortungsvollen Quellen
Paper from responsible sources
FSC® C105338
FSC
www.fsc.org

Printed by Books on Demand GmbH, Norderstedt / Germany